PUT

What role, if any, does formal logic play in characterizing epistemically rational belief? Traditionally, belief is seen in a binary way - either one believes a proposition, or one doesn't. Given this picture, it is attractive to impose certain deductive constraints on rational belief: that one's beliefs be logically consistent, and that one believe the logical consequences of one's beliefs. A less popular picture sees belief as a graded phenomenon. This picture (explored more by decision-theorists and philosophers of science than by mainstream epistemologists) invites the use of probabilistic coherence to constrain rational belief. But this latter project has often involved defining graded beliefs in terms of preferences, which may seem to change the subject away from epistemic rationality.

Putting Logic in its Place explores the relations between these two ways of seeing beliefs. It argues that the binary conception, although it fits nicely with much of our commonsense thought and talk about belief, cannot in the end support the traditional deductive constraints on rational belief. Binary beliefs that obeyed these constraints could not answer to anything like our intuitive notion of epistemic rationality, and would end up having to be divorced from central aspects of our cognitive, practical, and emotional lives.

But this does not mean that logic plays no role in rationality. Probabilistic coherence should be viewed as using standard logic to constrain rational graded belief. This probabilistic constraint helps explain the appeal of the traditional deductive constraints, and even underlies the force of rationally persuasive deductive arguments. Graded belief cannot be defined in terms of preferences. But probabilistic coherence may be defended without positing definitional connections between beliefs and preferences. Like the traditional deductive constraints, coherence is a logical ideal that humans cannot fully attain. Nevertheless, it furnishes a compelling way of understanding a key dimension of epistemic rationality.

Putting Logic in its Place

Formal Constraints on Rational Belief

DAVID CHRISTENSEN
University of Vermont

Clarendon Press · Oxford

This book has been printed digitally and produced in a standard specification in order to ensure its continuing availability

OXFORD
UNIVERSITY PRESS

Great Clarendon Street, Oxford OX2 6DP

Oxford University Press is a department of the University of Oxford.
It furthers the University's objective of excellence in research, scholarship,
and education by publishing worldwide in

Oxford New York

Auckland Cape Town Dar es Salaam Hong Kong Karachi
Kuala Lumpur Madrid Melbourne Mexico City Nairobi
New Delhi Shanghai Taipei Toronto
With offices in
Argentina Austria Brazil Chile Czech Republic France Greece
Guatemala Hungary Italy Japan South Korea Poland Portugal
Singapore Switzerland Thailand Turkey Ukraine Vietnam

Oxford is a registered trade mark of Oxford University Press
in the UK and in certain other countries

Published in the United States
by Oxford University Press Inc., New York

Reprinted 2008

ISBN 978-0-19-920431-1

For my parents,
Roshan Bharucha Christensen
and
James Roger Christensen

PREFACE

WHEN people talk informally about belief, "rational" and "logical" are often used almost synonymously. And even those who think carefully and precisely about rational belief often take logic to play an important role in determining which beliefs are rational. Explaining the importance of logic to students, philosophers often say things like, "Rational beliefs must be logically consistent with one another," or "If you believe the premises of a valid argument, then, if you are rational, you must believe the conclusion." This book aims to show that logic does indeed play an important role in characterizing ideally rational belief, but that its role is quite different from what it is often assumed to be.

The first chapter sets up parameters for the book's approach: it will focus on epistemic (rather than pragmatic) rationality; it will look at conditions on simultaneous rational beliefs (rather than on rational changes of belief); and it will concentrate on global rationality conditions for an agent's whole system of beliefs (rather than on local conditions for the rationality of particular beliefs). These choices are designed to focus the inquiry where formal logic is most likely to be useful in characterizing ideal rationality.

The second chapter ties the book's central question to a choice between two basic conceptions of belief. The standard *binary* model sees belief as an all-or-nothing state: either you believe P, or you don't. The *graded* model sees belief as coming in degrees. The two conceptions invite very different formal rationality conditions. Rational binary beliefs are often held to be subject to *deductive cogency*, which requires that an agent's beliefs form a logically consistent set which includes all the logical consequences of what the agent believes. Graded beliefs are often held to be subject to *probabilistic coherence*: the requirement that they obey the axioms of probability theory. Chapter 2 argues that probability theory is best seen not as a new logic for graded belief, but as a way of applying standard deductive logic to graded belief. It explores different ways of understanding the relation between binary and graded belief, concluding that the way one sees this relation has important implications for the questions of whether and how beliefs are subject to formal rationality constraints.

The third and fourth chapters argue that ideally rational binary belief is not subject to deductive cogency. Chapter 3 begins with the "Preface Paradox," which poses a well-known challenge to deductive cogency requirements. The chapter examines and rejects attempts to avoid the problem by understanding cogency requirements in weakened ways. It then develops an extended version of a preface case which shows how the intuitively irrational beliefs required by deductive cogency will in certain cases cascade into massive irrationality. This highlights the problem's severity and illuminates what's absurd about the beliefs required by deductive cogency. The chapter then shows that situations with Preface Paradox structure occur commonly in ordinary life. Finally, it examines and rejects attempts to defend deductive constraints by explaining away our troublesome intuitions in preface cases and in related cases involving the "Lottery Paradox."

Chapter 4 takes on a deeper sort of response to preface and lottery cases. It has been argued that the fundamental purposes of binary belief require deductive cogency. The chapter examines and rejects several arguments of this sort, the strongest being that if deductive cogency were not rationally required, deductive arguments would have no rational force. The chapter develops and defends an alternative explanation of the epistemic importance of deductive arguments, rooted in a probabilistic coherence constraint on graded belief, and in the relation that rational binary belief would bear to rational graded belief on any plausible account. The chapter ends with a discussion of whether binary belief has any epistemic importance; it argues that although our binary way of talking and thinking about belief may be very useful, it may not in the end capture any important aspect of rationality. Moreover, further development of extended preface-type cases shows that the sort of binary belief that was subject to deductive cogency could not have the connections to central aspects of our practical reasoning, our assertions, and our emotions that seem to give belief its point.

The fifth chapter turns to the positive task of defending probabilistic coherence as a logical constraint on graded belief. The two main strands of argument for this view in the literature are Dutch Book and Representation Theorem arguments. Unfortunately, both of these sorts of argument seek to defend constraints on graded belief by positing very tight connections between graded beliefs and *preferences*, which are not clearly within the epistemic domain. In fact, proponents of these arguments—who have tended to be decision-theoretically oriented philosophers of science, rather than mainstream epistemologists—have typically sought to *define* beliefs in terms of preferences. This may seem to change the subject away from epistemology proper, which I suspect helps explain why probabilistic approaches to rationality have not found more support among mainstream epistemologists. The chapter argues that defining graded belief in terms of preferences requires an insupportable metaphysics of belief, and thus that the arguments as they have typically been offered fail. Nevertheless, the chapter shows that the arguments can be reworked in a way that employs intuitively plausible normative principles connecting preferences with beliefs, eliminating the need for positing implausible metaphysical or definitional connections. Thus, probabilistic coherence can be defended without making beliefs into something they are not.

The final chapter addresses the issue of idealization in epistemology. Some have argued that probabilistic coherence in particular imposes an unacceptably high level of idealization; others would even reject deductive cogency as imposing excessive idealization. The chapter examines several reasons that have been offered for rejecting formal models of rational belief on the basis of excessive idealization. It defends the interest of idealized formal models in thinking about rationality, arguing that the human unattainability of logical perfection does not undermine the normative force of logical ideals.

This book has benefited greatly from correspondence, informal discussion, and comments on drafts from many friends and colleagues. I would like to thank Sin yee Chan, Keith DeRose, Jim Joyce, Mark Kaplan, Hilary Kornblith, Arthur Kuflik, Don Loeb, Patrick Maher, Bill Mann, Mark Moyer, Dana Nelkin, Derk Pereboom, and Jonathan Vogel. I'm especially grateful to Kornblith and Pereboom, who read every part of the manuscript at least once, and provided invaluable help and encouragement throughout.

I'd also like to thank an anonymous reader for Oxford University Press, and Mark Kaplan again, in his capacity as a not-so-anonymous reader for Oxford, for very useful suggestions. I'm grateful to the ACLS and the University of Vermont for sabbatical support, and to Leslie Weiger for administrative support. Chapter 5 is based on two papers published earlier: "Dutch-Book Arguments Depragmatized: Epistemic Consistency for Partial Believers," *Journal of Philosophy* 93 (1996): 450–79, and "Preference-Based Arguments for Probabilism," *Philosophy of Science* 68 (2001): 356–76. I thank the publishers for permission to reprint this material here. Material from Chapters 3 and 4 was presented to the Dartmouth–UVM Philosophy conference, and I'd like to thank my commentator, Jim Moor, and the other participants in the conference for stimulating discussion.

Finally, I'd like to thank my wife, Ruth Horowitz, and my children, Sophie and Sam Horowitz, for their love, their support, and their unfailing knack for putting philosophical pursuits into proper perspective. I know this will come in handy when, as will inevitably occur, errors are found in this book.

DC

CONTENTS

1 LOGIC AND RATIONAL BELIEF

1.1 Logic and Reason

If there is one respect in which humans differ most fundamentally from the other animals, perhaps it is our superior ability to reason about, and understand, the world. The main product of our reasoning, and medium of our understanding, is, of course, also our chief representation of the world—our system of beliefs.

Two dimensions of evaluation come immediately to mind in evaluating a person's beliefs. The first, and most obvious, is accuracy. Beliefs can represent the world more or less accurately, it seems, and the more accurate they are, the better. But philosophers have long been interested in a distinct dimension of evaluation. Some beliefs are more *rational* than others. And though the dimensions of accuracy and rationality may well be linked, they are evidently not the same. A fool may hold a belief irrationally—as a result of a lucky guess, or wishful thinking—yet the belief might happen to be accurate. Conversely, a detective might hold a belief on the basis of careful and exhaustive examination of all the available relevant evidence—in a paradigmatically rational way—and yet the evidence might happen to be misleading, and the belief might turn out to be way off the mark.

The point of evaluating rationality, as well as accuracy, of beliefs surely has to do with our desire to assess the agent qua believer. In doing so, we try to abstract from a certain kind of luck, or accidentalness. The fool is no better a thinker for having guessed correctly. The detective is no poorer a thinker for having encountered misleading clues. Rational beliefs, it seems, are those arising from good thinking, whether or not that thinking was successful in latching on to the truth.

But what is it that makes thinking "good"? A standard answer to this question is that, at least in part, good thinking is logical thinking. Thus logic has been at the center of philosophical thought about rationality since the time of the ancient Greeks. And the discipline of logic as practiced today incorporates at its center exactly the division between good thinking and accuracy mentioned above. The first lesson in most logic courses—and in many general introductions to philosophy—distinguishes soundness from validity. The latter, which is often thought of as the central concern of the logician, abstracts away from issues of actual truth and falsity to concentrate on studying correct and incorrect relations between claims—relations that are thought to be embodied in good and bad thinking, respectively. Of course, there may turn out to be a deep connection between considerations of truth and considerations which make certain relations between claims "logical." But the first concern of the logic teacher is typically to separate logical relations from factual ones.

The motivations for studying the logic of rational belief are undoubtedly various. We might seek to improve the thinking of others, or of ourselves, by providing rules that people could self-consciously employ in forming or revising their beliefs. We might seek to diagnose confusion in cases where our thinking naturally leads us to paradoxical results. Or we might simply seek a purely theoretical understanding of good thinking, for its own sake. But regardless of motivation, something like the following idea seems to be presupposed in studying logic: that the correct logic will provide a model for ideally rational belief. It is this idea that I would like to examine and, eventually, defend.

'Logic,' of course, is not a term that is used consistently, even within academic philosophy. Texts on logic discuss issues ranging from Gödel's incompleteness theorem to ways of identifying *ad hominem* arguments. What I have in mind is *formal* logic. Of course, even the meaning of 'formal' is not clear. Texts often refer to, e.g., entailments whose validity depends on logical form or structure; but in explaining what counts as form, the texts typically resort to examples. Thus, the fact that the sentence

(1) Sulfur is yellow

entails the sentence

(2) Sulfur is yellow or sulphur is green

might be cited as flowing from the fact that sentences of the form "P or Q" are always entailed by sentences of the form "P." This is a paradigmatic principle of formal logic. However, the claim that (1) entails

(3) Sulfur is not red

might well not be considered to flow from any fact about logical form. This is so even though one might claim that sentences of the form "x is not red" are always true when sentences of the form "x is yellow" are. And most would reject the claim that

(4) This rock is made of sulfur

entails

(5) This rock is made of the element with atomic number 16

as a matter of logic, even though many would hold that sentences of the form "x is made of sulfur" can only be true at possible worlds where sentences of the form "x is made of the element with atomic number 16" are true as well.

I do not propose here to settle the questions of whether the second of the above-mentioned entailments is a logical one, or why the third is not. Provisionally, let us take formal logic as concerned with at least the forms or structures created by the standardly accepted logical words such as 'not,' 'or,' 'and,' 'if…then,' 'all,' and 'some.' Could the correct logic of such forms (which is perhaps not the whole of formal logic) provide a model (undoubtedly a partial model) for ideally rational belief?

Before beginning to answer this question, it will be useful to clarify what sort of rationality is at issue, in several respects.

1.2 Pragmatic vs Epistemic Rationality

A distinction is often made between two senses of rationality, both of which can be applied to beliefs. The distinction is most easily illustrated with a touched-up version of Pascal's Wager. Suppose that, given the evidence available to me, it's unlikely that God exists. However, suppose that the evidence also makes it very likely that if God does exist it will be overwhelmingly in my best interests to toe the theistic line—not only in my actions, but in my beliefs. It could then be rational for me, in the pragmatic sense, to believe in God: given what I want, having that belief could be expected to be most advantageous relative to my ends.[1] But there is also a clear sense in which a belief adopted counter to the evidence would not be a rational one. It is this second, epistemic, sense of rationality that I am concerned with here.

This is not to deny that the two forms of rationality are connected. One might suggest that epistemically rational beliefs are those that would maximize one's expectation of reaching one's particularly epistemic goal or goals (such as believing true things or failing to believe false things). As it stands, the suggestion seems clearly wrong. After all, one can imagine a variant on the Pascal example in which the rewards for the counter-evidential belief were purely epistemic. Consider, for example, a case in which three-fingered aliens offer one vast new insight into physics,

[1] I should note that pragmatic rationality need not be tied to self-interest. It would have to be specified that, *given all my values*, including, e.g., any value I placed on my believing the truth and any value I (perhaps altruistically) placed on other things that could be affected by my beliefs, the expected value of my believing would be higher.

including both new information and corrections of many current misconceptions, on the condition that one believe that Ghengis Khan had three fingers (I'm assuming that the aliens' offer gives one no evidence for this proposition). In this case, it seems clear to me that if one somehow comes to believe that Ghengis Khan was three-fingered, this belief is epistemically irrational—whether or not the aliens can rationally be expected to come through with their part of the bargain. But there are more sophisticated ways of connecting pragmatic and epistemic rationality, some of which will be discussed in later chapters. At this point, I just want to distinguish our epistemic notion from the pragmatic one.

1.3 Diachronic vs Synchronic Rationality

Logic books are often written as though their central topic were inference. Arguments are set out in premise–conclusion form, and it is suggested that the premises represent an agent's present beliefs, and the conclusion a further belief that the agent should, after going through the argument, adopt. This suggests that the sort of rationality being addressed through logic is *diachronic* rationality. Diachronic rationality constrains the way beliefs are changed (or maintained) across time. The idea seems to be something like this: logic's rules of inference tell you which new beliefs you should adopt, on the basis of your current beliefs. Thus Modus Ponens tells you that if you believe P, and you also believe $(P \supset Q)$, then if you don't already believe Q, you should adopt the belief that Q.

It is well-known, of course, that this picture is too simple. After all, if you learn by the above logical argument that your beliefs entail Q, you might well want to revise your current belief that P, or your belief that $(P \supset Q)$, rather than adopt the new belief that Q. This is especially clear if you happen already to believe not-Q. But even if you're currently agnostic on Q, Modus Ponens itself provides no reason for preferring, e.g., becoming a Q-believer to becoming a P-agnostic.[2] In each of these cases, logic gives you no guidance at all regarding which option for revising your beliefs is preferable. Thus the "rules of inference" given in logic books cannot be thought of in any straightforward way as rules of diachronic rationality.[3]

In fact, despite the way logic texts often present the subject, it is far from clear that the basic concern of logic is with change of belief. A valid argument, as most often defined, is one in which it is impossible for the premises to be true while the conclusion is false. In other words, the notion of valid argument flows from a deeper notion, a notion involving the possibility of sentences simultaneously having certain truth-values. Facts about the possible truth-value distributions among the members of a certain set of sentences are not diachronic facts about those sentences. Of course, the facts about possible truth-value distributions have implications for inferences constructed from those sentences—e.g. that certain inferences are truth-preserving. But these implications do not really go beyond the fundamental facts about simultaneous truth-value distributions.

The view of logic as concerned most basically with simultaneous truth-value distributions suggests simpler ways of applying logic to the theory of rational belief. We saw above that logic did not help the rational agent choose between, on the one hand, adopting the

[2] I should note that, as Mark Moyer pointed out to me, in a typical case where you believe P and (P ⊃ Q) but are agnostic on Q, you will have evidence for the first two beliefs, and this might well give you good reason for adopting Q over abandoning P. But there are certainly cases in which abandoning P is the rational choice. For example, suppose that the general practice has been for Sophie to wash the dinner dishes. But this morning, I saw Sophie and Sam decide that tonight's dishwashing duties would be decided by flip of a coin: Sophie will wash iff the coin comes up heads. Nevertheless, not having this memory at the forefront of my mind, I now believe that Sophie will wash the dishes tonight. I also believe (though I don't bring this belief to mind at the same time as the last one) that Sophie will wash the dishes tonight only if the coin will come up heads. So I believe S and I believe (S ⊃ H), but I haven't put two and two together, and I don't believe H. In this case, when I do notice that my beliefs fail to respect Modus Ponens, it seems that I should become an S-agnostic rather than beginning to believe in H.

[3] Harman (1970, 1986) makes this point nicely. In the latter work, he draws the more dramatic conclusion that "there is no clearly significant way in which *logic* is specially relevant to reasoning" (1986, 20).

belief that Q and, on the other hand, dropping the belief that P or the belief that (P ⊃ Q). But logic did suggest some constraints on our agent's beliefs. In fact, the apparent need for *some revision or other* stemmed from the idea that there was something wrong with certain sets of *simultaneous* beliefs. Most plausibly, it may be claimed that logic precludes the option of rationally believing all of P, (P ⊃ Q), and not-Q at the same time. And one might also see logic as precluding the option of rationally remaining agnostic on Q while believing both P and (P ⊃ Q). Generalizing these two suggestions yields the two most prominent proposals for using logic to constrain rational belief: the requirement that a rational agent's beliefs be *logically consistent*, and the requirement that the rational agent's beliefs be *closed under deduction* (i.e. that the agent believe everything logically implied by her beliefs). Both of these proposals are, of course, for synchronic constraints on rationality. Thus I would like to concentrate on the question of whether logic can provide synchronic constraints on ideally rational beliefs.

This may seem misguided, if we want rational beliefs to be those that are in some sense the products of good thinking. "Good thinking" seems to be an activity, and a notion of rationality that tried to capture what good thinking was might seem to be an essentially diachronic notion, involving the evaluation of how we change (or maintain) beliefs through time. Now some have argued that epistemic rationality is fundamentally independent of diachronic considerations.[4] But even if we put this issue aside, it seems clear that synchronic constraints could play a role in a notion of rationality that was tied to thinking well. For example, "maintain logical consistency" is certainly a rule for how to think, but the constraint it places on ideally rational beliefs concerns how beliefs at a given time relate to other beliefs at that same time. Thus the importance of synchronic constraints is perfectly compatible with conceptions of epistemic rationality according to which it has important dynamic dimensions.

[4] See e.g. Foley (1993).

1.4 Local vs Global Rationality

One of the first things that is intuitively apparent when one begins to think about rational belief is that some of an agent's beliefs may be more rational than others. For example, although I may be highly responsive to the evidence when I form beliefs about the character, talents, and physical attractiveness of most children I happen to meet, my beliefs about how my own children fare along these dimensions may be subject to considerable non-evidential influences. Clearly, a complete study of rationality must include those factors that affect the rationality of an agent's beliefs differentially. And of course, much work in epistemology has focused on describing what it is about an individual belief that makes it rational, or, more often, what makes it justified, or an instance of knowledge.

This approach encounters a well-known difficulty when it examines questions relating to how structural or logical factors influence epistemic merit. If we are asking, for example, whether my belief that Q is made rational in part by its following logically from my beliefs that P and $(P \supset Q)$, the answer to our question will in part depend on the epistemic status of the latter beliefs. In general, the merits (or demerits) a belief receives in virtue of its logical connections to other beliefs will depend on two entirely separate factors: the nature of the structural relations themselves, and the epistemic credentials of the related beliefs. And evaluating the second component incurs obvious regress problems, since the epistemic credentials of the related beliefs will depend on their structural relations to still other beliefs, whose own epistemic credentials then become relevant; and so on.

There are, of course, various solutions to the regress problem, and this is not the place to examine them. But we may note that focusing on structure leads us to see the rationality of an individual belief as tied into a complex web of logical relations. In short, there's a pressure toward the sort of holism that takes the entirety of a person's belief system as at least potentially relevant to the epistemic status of any given belief. I don't want to argue here that this degree of holism is correct. But it does seem clear that, insofar as logical relations are important to a belief's epistemic status, they are going to involve large sets of interrelated beliefs.

This is not to say that even a strong form of holism would preclude our allowing differences in epistemic merit to exist among the beliefs of a single agent. But it does make the task of accounting for these differences more complex. And the notions invoked in such an account will naturally be notions of degree—e.g. the degree of a given belief's coherence with other beliefs, or the directness of the connections between the given belief and relevant other beliefs—notions that are unlikely to find precise formal characterization. Thus, insofar as we can capture the factors that are responsible for the differing degrees of rationality enjoyed by various of an agent's individual beliefs, they will likely involve complex and non-formal notions.

This suggests that in studying formal constraints on rationality we might best begin by focusing not on differences in epistemic status among individual beliefs, but rather on the possibility of giving formal constraints on the whole set of an agent's beliefs. And indeed, the initially attractive formal constraints mentioned above—deductive consistency and deductive closure—are global constraints of just this sort. The same holds for probabilistic formal constraints on degrees of belief that have been advanced by many writers.

This global approach to formal constraints on rationality dovetails nicely with seeing such constraints as aspects of an epistemic *ideal*. Concentrating on ideal rationality allows us to put aside, for the moment at least, the difficult questions involved in balancing those non-ideal factors responsible for differences in rationality among the beliefs of a realistic agent.

This approach does not, of course, involve any claim to the effect that all departures from ideal rationality are on a par. For instance, in his coherence-based account of empirical justification, Laurence

BonJour holds that an inconsistency anywhere in an agent's beliefs diminishes, to some extent, the justification of every one of her beliefs (since no belief can cohere perfectly with the entire corpus). BonJour offers no detailed way of assessing how much the rationality of the various beliefs in an agent's corpus is affected by different sorts of inconsistency, but he suggests that different sorts of inconsistency would diminish justification to different degrees. Still, BonJour holds that ideal justification requires that the whole set of the agent's beliefs be logically consistent.[5]

Thus it should be clear that, in concentrating on ideal rationality, we are deliberately putting aside important questions. Real people will presumably always fall below the ideal standards, and any complete account of epistemic rationality must eventually give us insight into the degrees of sub-ideal rationality we find in real people's beliefs. But the fact that a correct description of ideal rationality would not exhaust epistemology does not vitiate the interest of such a description.

Nevertheless, this facile observation—that one cannot do everything at once in epistemology—leaves open some serious worries about both the degree and the nature of idealizations involved in proposals for imposing formal constraints on rational belief. Some of these worries are specific to particular formal models, while others involve the very notion that rationality can be tied to logical standards which are out of reach for real people. These questions will be explored in detail in later chapters. For now, I only want to claim that examining global conditions on ideal rationality is a promising place to start in studying the question of what role formal models can play in understanding rational belief.

To sum up: Our main question is whether formal models have an important role to play in understanding rational belief. This question is complicated by the fact that there are many aspects and varieties of rationality that might be thought to apply to belief, and there can be different approaches to studying a given aspect or

[5] See e.g. BonJour's "Replies and Clarifications," in Bender (1989, 284).

variety of belief-rationality. My aim has been to focus on that part of the theory of rational belief in which formal models seem most likely to play a useful part. Let us begin, then, by examining formal constraints that apply to the whole set of beliefs an agent has at a given time, and asking whether such constraints can provide a model for ideal epistemic rationality.

2 TWO MODELS OF BELIEF

2.1 Models of Belief and Models of Rationality

WHEN people talk about the world, they typically make unqualified assertions. In ordinary contexts, it is natural to take those assertions as reflecting beliefs of the speaker; if a speaker says "Jocko cheated" (as opposed to "Jocko probably cheated," or "Jocko must have cheated"), we infer that she bears a fairly simple relation to the claim that Jocko cheated—she believes it. This relation often does not seem to be a matter of degree; either one believes that Jocko cheated, or one doesn't.[1]

Similarly, when people talk explicitly about their beliefs, they often seem to presuppose an all-or-nothing notion. Questions such as "Do you believe that Jocko cheated?" oftentimes seem unproblematically precise. The model of belief that seems implicit in these cases is black-and-white: belief is an attitude that one can either take, or fail to take, with respect to a given claim.

Of course, one also may disbelieve a claim, which is clearly a different thing from failing to believe it, despite the fact that it is natural to express, e.g., disbelief in the claim that Jocko cheated by saying "I don't believe that Jocko cheated." But disbelief need not be seen as a third attitude that one can take to a claim. Disbelieving a claim is naturally understood as believing the claim's negation. Failing to believe either a claim or its negation seems naturally to be expressed by assertions such as "I don't know whether Jocko

[1] Unqualified assertions may indicate something more than belief, such as claims to knowledge. (One may react to a challenge to one's unqualified assertion by saying, e.g., "Well, I *believe* that Jocko cheated.") But even on this stronger reading of what assertions indicate, they seem to indicate a state that includes an all-or-nothing state of belief.

cheated." So the model of belief that seems implicit in much ordinary thought is naturally taken to be a binary one.[2]

A binary model of belief also fits in very naturally with philosophical analyses of knowledge. Knowledge has typically been seen as belief-plus-certain-other-things. The belief part has typically been taken as unproblematic—either the agent believes the claim or she doesn't—and the main task of the theory of knowledge has been taken to be that of providing an adequate specification of what, besides belief, knowledge requires. Even those epistemologists who concentrate on the justification of belief—a topic close to our own—have often seen justification as one of the things a belief needs in order to count as knowledge. Thus mainstream epistemologists of various persuasions have typically employed a binary model of belief.

Nevertheless, the binary model does not provide the only plausible way of conceiving of belief. It is clear, after all, that we have much more confidence in some things we believe than in others. Sometimes our level of confidence in the truth of a given claim decreases gradually—say, as slight bits of counterevidence trickle in. As this occurs, we become less and less likely to assert in an unqualified way (or to say unqualifiedly that we believe) the claim in question. But reflection on such cases fails to reveal any obvious point at which belief suddenly vanishes. At no time does there seem to occur a crisp qualitative shift in our epistemic attitude toward the claim. This suggests that underlying our binary way of talking about belief is an epistemic phenomenon that admits of degrees.

Degrees of belief reveal themselves in numerous ways other than in our introspection of different levels of confidence. Famously, in confronting practical problems in life, whether about what odds to bet at or about whether to carry an umbrella when leaving the

[2] One might quite reasonably want to avoid equating disbelief in P with belief in P's negation. In that case, one would naturally see discrete belief as a trinary notion, encompassing three distinct attitudes one might take toward a proposition: belief, disbelief, and withholding judgment. Since nothing relevant to the present discussion turns on the difference between these ways of understanding discrete belief, I will continue to speak of the "binary" conception.

house, our decisions and actions seem to be explained by degrees of belief. Rational explanations of an agent's actions typically make reference to the agent's beliefs and desires. The desire-components of such explanations obviously depend not only on the contents of the agent's desires, but on their strengths. And similarly, the belief-components of such explanations depend on the agent's degrees of confidence that the various possible choices open to her will lead to outcomes she cares about. The common-sense psychological principle that underlies these explanations seems to be a rough approximation of expected utility maximization: in the textbook umbrella case, for example, the greater an agent's confidence that leaving the umbrella at home will result in her getting wet, and the more strongly she disvalues getting wet, the less likely she will be to leave the umbrella at home. Thus, a sizable minority of epistemologists have approached the rationality of belief from a perspective closely intertwined with decision theory, a perspective in which degrees of belief are taken as fundamental.

Both the binary and the graded conceptions of belief enjoy, I think, at least a strong prima facie plausibility. And each conception figures in apparently important philosophical thought about rationality. Thus, although it could turn out in the end that one (or both) of these conceptions failed to pick out any epistemically important phenomenon, we should not dismiss either one at the outset as a potential home for formal rationality requirements. Still, this leaves open a number of possible approaches to the objects of epistemic rationality. One might see binary belief as reducing to graded belief, or graded belief to binary belief. In such a picture, there would be at bottom only one fundamental object of rational appraisal. Alternatively, one might see two independent (though undoubtedly related) epistemic phenomena. In this case, perhaps each would be answerable to its own distinctive set of rational demands.

Getting clear on this issue is important to our purposes, because the two conceptions of belief seem to invite quite different kinds of formal models. The traditional binary conception of belief meshes naturally with straightforward applications of deductive logic. On the binary conception, there is a set of claims that a given agent believes. The basic idea is, roughly, that membership in this set of claims ought (ideally) to be conditioned by the logical properties of, and relationships among, those claims. As we've seen, deductive consistency and deductive closure are prominent candidates for constraints on an ideally rational agent's set of binary beliefs.

By contrast, the graded conception of belief requires quite a different treatment. On this conception, there is not one distinctive set of claims the agent "believes"; instead, the agent takes a whole range of attitudes toward claims. At one end of the spectrum are those claims the agent is absolutely certain are true, at the other end are claims the agent is absolutely certain are false, and in between are ranged the vast majority of ordinary claims, in whose truth the agent has intermediate degrees of confidence. The standard formal models for ideally rational degrees of belief involve using the probability calculus. Degrees of belief are taken to be measurable on a scale from 1 (certainty that the claim is true) to 0 (certainty that the claim is false). An ideally rational agent's degrees of belief must then obey the laws of probability; to use the common terminology, they must be probabilistically coherent.

The probability calculus is often referred to as a logic for degrees of belief. It might be more illuminating to see it as a way of applying standard logic to beliefs, when beliefs are seen as graded. The constraints that probabilistic coherence puts on degrees of belief flow directly from the standard logical properties of the believed claims. Consider, for example, the fact that probabilistic coherence requires one to believe ($P \vee Q$) at least as strongly as one believes P. This flows directly from the fact that ($P \vee Q$) is logically entailed by P. In fact, we can plainly see connections between the natural ways logic has been taken to constrain belief on the binary and graded conceptions. The dictate of logical closure for binary beliefs requires that

> an ideally rational agent does not believe P while failing to believe ($P \vee Q$).

Probabilistic coherence of graded belief requires that

> an ideally rational agent does not believe P to a given degree while failing to believe (P ∨ Q) to at least as great a degree.

Similarly, logical consistency of binary belief requires that

> an ideally rational agent does not believe both P and ∼ (P ∨ Q);

in other words, if she believes one of the two sentences, she does not believe the other. Probabilistic coherence of graded belief requires that

> an ideally rational agent's degrees of belief in P and ∼(P ∨ Q) do not sum to more than 1;

in other words, the more strongly she believes one of the two sentences, the less strongly she may believe the other.

The idea that the probability calculus functions less as a new logic for graded belief than as a way of applying our old logic to graded belief may be supported by looking at the basic axioms of the probability calculus. Put informally, they are as follows (where pr(P) stands for the probability of P):

(1) For every P, pr(P) $\geq$ 0.
(2) If P is a tautology, then pr(P) = 1.
(3) If P and Q are mutually exclusive, then pr(P ∨ Q) = pr(P) + pr(Q).

The above formulation is quite typical in using the notions of tautology and mutual exclusivity. These notions are, of course, the standard logical ones. Presentations of the second axiom sometimes use "necessary" rather than "a tautology," but insofar as necessity and logical truth come apart, it is the latter that must be intended. No one thinks, presumably, that the axioms of probability should be applied to rational belief in a way that requires "Cicero is Tully" to have probability 1.

This observation suggests that the import of the standard axioms is parasitic on a pre-understood system of deductive logic. On any system of logic, (P & Q) will entail P, and this will be reflected directly in restrictions on probabilistically coherent degrees of beliefs that one may have in these propositions. But the boundaries of logic are not entirely obvious. If it is a matter of logic that □P entails P, then (□P ⊃ P) will be a tautology, and □P and ~P will be mutually exclusive, and this will in part determine which degrees of belief involving these sentences can be probabilistically coherent. Similarly, when we decide whether, as a matter of logic, □P entails □□P, or "x is yellow" entails "x is not red," or "x is made of sulfur" entails "x is made of the element with atomic number 16," we will thereby determine the contours of probabilistic coherence. That is why the axioms of probability are better seen not as a distinct logic for graded beliefs. The probability calculus is most naturally seen as just giving us a way of seeing how rational graded beliefs might be subject to formal constraints derived directly from the standard logical structures of the relevant propositions.

Now it is true that there are ways of axiomatizing the probability calculus that do not separate the probabilistic axioms from those of deductive logic. For example, Karl Popper (1959) gives an axiomatization for conditional probability that incorporates standard propositional logic (he intends it as a generalization of deductive propositional logic). Hartry Field (1977) extends Popper's technique to give an axiomatization that incorporates predicate logic (Field intends not to generalize deductive logic, but rather to provide a truth-independent semantics which reflects conceptual roles rather than referential relations).[3] We should be careful, then, about what we conclude from examining standard formulations of probability theory (or the formulations used by the theory's developers): even if the standard axiomatizations are intuitively natural, that does not prove that the probability calculus is, at the most fundamental level, parasitic on a conceptually prior system of deductive logic.

[3] The relevance of this point was brought to my attention by a referee. See also Hawthorne (1998) for further development and related references.

However, the point remains that probability theory is in no way independent of the ordinary logical relations familiar from deductive logic—relations that derive from important structural patterns involving 'and,' 'not,' 'all,' etc. The constraints that any version of probability theory places on degrees of belief flow from exactly these patterns. And the standard way of axiomatizing probability shows that, for any of the familiar notions of deductive consistency, there will be a probabilistic way of taking account of that logic's structural basis.

For both models of belief, then, the prominent proposals for imposing formal constraints on ideal rationality are rooted in logic. But the logic-based constraints take quite different forms for the different models of belief. Moreover, it turns out that the arguments both for and against the imposition of the formal constraints are quite different for binary and graded belief. Thus our examination of the plausibility of formal constraints on rational belief will clearly be shaped by our choice of how to see rational belief itself.

2.2 Unification Accounts

We saw above that both conceptions of belief enjoy enough plausibility to be worth exploring, and thus that we should not reject either out of hand. But even putting aside the eliminationist option of rejecting one of the conceptions as not picking out any real phenomenon, one might favor what might be called a unification approach. One might hold that one sort of belief was really only a special case or species of the other. If such a view were correct, it clearly could help determine our approach to formal rationality.

Perhaps the less attractive unificationist option is to take graded beliefs as nothing over and above certain binary beliefs. Let us consider an example in which a graded-belief description would say that an agent had a moderate degree of belief—say, 0.4—in the proposition that Jocko cheated on Friday's test. Should we see this graded belief as really consisting merely in the agent's having some particular binary belief? If so, we should presumably turn our attention straightforwardly to deductive constraints.

The problem with this proposal stems from the difficulty of finding an appropriate content for the relevant binary belief. A first try might be that the probability of Jocko's having cheated on Friday's test is 0.4. But what does "probability" mean here? The term is notoriously subject to widely divergent interpretations. Some of these interpretations—those of the "subjectivist" variety—define probability explicitly in terms of graded belief. Clearly, if graded beliefs are merely binary beliefs about probabilities, the probabilities involved must not be understood this way.

On the other hand, if we understand probabilities in some more objective way, we risk attributing to the agent a belief about matters too far removed from the apparent subject matter of her belief. For example, if probabilities are given a frequency interpretation, we will interpret our agent as believing something like: Within a certain specific reference class (cases where people had a chance to cheat on a test? cases where people like Jocko had a chance to cheat on a test? cases where Jocko himself had a chance to cheat on a test on a Friday? ...), cheating took place in 4/10 of the cases. Yet it is hard to believe that any thought about reference classes need even implicitly be present in the mind of an agent to whom we would attribute a 0.4 degree of belief in Jocko's having cheated. If probability is given a propensity interpretation, things are no better. Since the belief in question is about a past event, we cannot say that the agent believes that some current setup is disposed to a certain degree to end up with Jocko cheating on the test in question. And it seems quite implausible to analyze our agent's belief as really being about the way Jocko was disposed to behave at a certain point just prior to the test.

One could object to this argument that precise degrees of belief are almost never correctly attributable, and that my example therefore should not have specified a degree as specific as 0.4 in the first place. The agent, it might be held, really only harbored a (binary) belief that Jocko's cheating was quite possible, but not highly probable. But while there may be some point behind the charge that the attribution of precisely a 0.4 degree of belief in this case is unrealistic, softening the focus here to talk about more vague probability-beliefs does not address the present worry. The worry, after all, was that when people have intermediate degrees of belief in propositions, they need not have any beliefs at all about, e.g., frequencies within reference classes, or propensities.

Of course, these examples are based on quick and crude caricatures of prominent objective interpretations of probability, and still other objective accounts of probability do exist. But for our purposes, these examples serve well enough to show how unnatural it is to identify an agent's having a certain degree of confidence in a particular proposition with that agent's having an all-or-nothing belief about some non-belief-related proposition about objective probabilities.

Moreover, it is clear that, in general, people's attitudes do come in degrees of strength. Presumably, no one would doubt the existence of degrees of strength with respect to people's hopes, or fears, or attractions, or aversions. Yet on the unification view about belief that we have been considering, strength of confidence would have no reality independent of (binary) beliefs about objective probabilities. I see little reason to accept such a view. So although this sort of unification would simplify matters by turning our attention to deductive, as opposed to probabilistic, constraints on rational belief, it seems unlikely that trying to simplify matters in this way would be successful.

A more promising sort of unification would work in the opposite way. We might see binary belief as a special case or species of graded belief: one would believe something in the binary sense if she believed it (in the graded sense) with a strength that met a certain threshold. Two variants of this proposal have in fact been advanced. According to one, binary belief is identified with graded belief of the highest degree (1); on this account, to believe P is to be certain that P. According to the other account, the threshold is lower (and may not be precisely specified); on this account, to believe P is to be sufficiently confident, but not necessarily certain, that P. Let us consider these accounts in turn.

The certainty proposal is, I think, less plausible. If the binary conception of belief derives its plausibility from our habit of making unqualified assertions, and from our ordinary ways of thinking and talking about belief, then the plausible notion of binary belief is of an attitude that falls far short of absolute certainty. We often assert, or say that we believe, all kinds of things of which we are not absolutely certain. This is particularly clear if the plausibility of the graded conception of belief is rooted in part in how belief informs practical decision. Insofar as degree of belief is correlated with practical decision-making, the highest degree of belief in P is correlated with making decisions that completely dismiss even the tiniest chance of P's falsity. For example, having degree of belief 1 in Jocko's having cheated would correlate with being willing literally to bet one's life on Jocko's having cheated, even for a trivial reward. Surely this level of certainty is not expressed by ordinary unqualified assertions; nor is it what we usually want to indicate about ourselves when we say, e.g., "I believe that Jocko cheated," or what we want to indicate about others when we say, e.g., "Yolanda believes that Jocko cheated."

Now one might resist taking too strictly our everyday tendencies to attribute belief in cases such as Jocko's cheating, and still insist that there is an important class of ordinary propositions about the external world which we rationally accord probability 1. Isaac Levi (1991) has argued that we do, and should, have this sort of "full belief" even in propositions that we come to believe by methods which, we recognize, are not absolutely reliable. When we accept such propositions as evidence, we "add [them] to the body of settled assumptions," which are "taken for granted as settled and beyond reasonable doubt" (1991, 1). According to Levi, these propositions then function as our standard for "serious" (as opposed to merely logical) possibility.

However, it does not seem to me that we are actually fully certain even of the things we typically take for granted or treat as evidence. It is, of course, true that there are many propositions which, in some rough sense, we regard as settled in our practical and theoretical deliberations. For example, scientists studying the effects of a new drug on rats may accept as evidence a proposition such as

> The rats treated with drug D died, while the rats in the control group lived.

In evaluating hypotheses about the drug, the researchers will consider various explanations for this evidence—that drug D caused the deaths of the treated rats; that the batch of saline solution in which drug D was dissolved contained a contaminant that caused the deaths of the treated rats; that it was just a coincidence; etc. But they will not consider the possibility that the evidence proposition is actually false. In an ordinary sense, this possibility will not be taken as "serious."

Does this mean that the researchers are absolutely certain of the evidential proposition? I don't think so. We would not, for example, expect one of them to be willing to bet the lives of his children against a cup of coffee on the proposition's truth. And we would not think that it would be reasonable for him to do this. Why? Because there is some incredibly small chance that, e.g., the lab technician switched the rats around to make the experiment "come out right." What would explain the researcher's reluctance to take the bet (or our reluctance to call the bet reasonable) is precisely the fact that the researcher is not completely certain of the evidential proposition.

But let us put this sort of doubt aside, and consider the consequences of accepting a unification account on which binary belief was identified with graded belief of probability 1. It remains true that the graded conception of belief has within it the notion of "full belief," or belief with degree 1. And one might argue for a kind of unification (perhaps one that deviated from some aspects of our intuitive conceptions) by identifying binary belief with full belief.

If we were to accept this sort of unification, what impact would it have on the question of formal constraints on rational belief?

Clearly, the fundamental approach to rational constraints would be the one appropriate to graded belief—presumably, a probabilistic one. And adopting such an approach would actually automatically impose constraints on binary belief—in fact, constraints that would at least come close to the traditional deductive constraints of consistency and closure.[4] But the status of the (approximation to the) traditional deductive constraints on this picture would be derivative. Insofar as the certainty proposal is plausible, then, it argues for taking a probabilistic approach to formally constraining rational belief.

Perhaps, however, it is more plausible to unify the two conceptions of belief by setting the binary belief threshold at some level below that of certainty. One needn't hold that our ordinary notion picks out some precise cutoff value ("if it's believed to at least degree 0.9, it is Believed"); one might hold instead that the border of binary belief is a vague one. Still, one might develop a model of rational belief that incorporated a precise (if somewhat arbitrary) cutoff point, in order to study the formal constraints that might apply on any such precisification.

This sort of unification comes closer than does the certainty proposal to fitting with our ordinary practices of unqualified assertion and belief-attribution. By and large, it seems, we do make assertions and attribute (binary) beliefs in cases where degrees of

[4] The constraints imposed on full beliefs by the probability calculus coincide with those imposed on binary beliefs by traditional consistency and closure conditions in many ways. For example, one cannot fully believe a contradiction; one must fully believe tautologies; one cannot have less than full belief in $(P \vee Q)$ while having full belief in P; and one cannot have full belief in all of P, $(P \supset Q)$, and $\sim Q$. The divergences can occur in certain contexts involving infinite sets of beliefs. For example, if one is certain that something is located at a point somewhere in a given area, but thinks that all the infinite number of points in the area are equally likely, it turns out that the probability assigned to the thing being at any one point must be 0, and hence the probability of it not being at that point must be 1. Thus one must have full belief that the thing is not at p, for each point p in the area—even though one also has full belief that the thing is at one of these points. In this sort of case, then, one has an inconsistent (though not finitely inconsistent) set of beliefs. See Maher (1993, ch. 6.2) for detailed discussion of this matter.

belief are fairly high. Thus, of all the unification proposals considered so far, this one may be the most likely to be correct.[5]

On this sub-certainty threshold account, it is not true that imposing probabilistic constraints on graded belief automatically imposes deductive-style constraints on binary belief. There's no reason to think, for example, that the set of things a rational agent believes to at least degree 0.9 should be consistent with one another. In fact, quite the reverse is true, for any sub-certainty threshold, as is made clear by lottery examples. (Consider a rational agent who has excellent evidence, and is thus very highly confident (> 0.999), that a particular 1,000-ticket lottery is fair, and that one of its tickets will win. For each ticket, his confidence that it won't win is 0.999. Thus he is rationally confident, to an extremely high degree, of each member of an inconsistent set of propositions.) Henry Kyburg famously used this point in arguing against taking deductive consistency to be a requirement on binary belief.[6] Others have used it in the opposite way, arguing that since deductive consistency is a constraint on binary belief, binary belief in a proposition cannot simply be a matter of having sufficient confidence in it.[7]

I don't want to take a stand here on whether our ordinary binary conception of belief is best understood as referring to a certain level of confidence. Although our assertion and attribution practices may fit better with this account than with the certainty account, the fit is not perfect, especially in lottery cases.[8] Still, one might well maintain that our talk of binary belief is most plausibly construed as referring to a high level of graded belief, and then work to explain away tensions with our assertion and attribution practice (e.g. by invoking principles of conversational implicature). How would such an approach affect the question of formal epistemic constraints?

[5] Foley (1993, ch. 4) provides a clear and detailed defense of this sort of view.

[6] See his "Conjunctivitis," in Swain (1970).

[7] For recent examples of this argument, see Maher (1993, ch. 6) and Kaplan (1996, ch. 3).

[8] The status of our attitudes toward lottery tickets (and related matters) will be discussed in more detail in later chapters.

As noted already, the classical constraint of deductive consistency for binary beliefs would have to be given up. The same would then hold for deductive closure: in standard lottery cases, for example, "no ticket will win" follows deductively from propositions each of which meets the confidence threshold for belief, but it does not come close to meeting that threshold itself. As Kyburg points out, binary belief on such an account could still obey vastly weakened versions of these constraints. Beliefs could obey the "Weak Consistency Principle" requiring that no one belief was a self-contradiction. And they could respect a weak version of deductive closure, the "Weak Deduction Principle," requiring that anything entailed by a single belief was also believed.

Nevertheless, for our purposes, the important point is that these weak principles are simply automatic consequences of imposing probabilistic coherence on the agent's graded beliefs. Weak Consistency would follow from probabilistic coherence because contradictions have probability 0, and thus would fall below the threshold. Weak Deduction would follow because any logical consequence of a sentence must have at least as high a probability, so if P meets the threshold and P entails Q, Q must meet the threshold as well.

In fact, Kyburg points out that somewhat stronger consistency principles can be imposed, depending on the threshold chosen. If the threshold is over 0.5, "Pairwise Consistency" follows: no pair of inconsistent propositions may be believed (though an inconsistent triad is not ruled out). And in general, as the threshold for belief becomes higher, increasingly larger sets of jointly inconsistent beliefs will be prohibited. Of course, even at a very high threshold (e.g. 0.99), the system will allow large sets (e.g. 101) of jointly inconsistent beliefs.[9]

9 Think of an agent who is extremely confident that a certain 100-ticket lottery is fair; the inconsistent set of beliefs will be 100 particular beliefs of the form "ticket n won't win," along with the general belief that one of the tickets will win. See Hawthorne and Bovens (1999) for an interesting and detailed exploration of the sorts of consistency constraints that may be imposed in lottery and related cases, given a threshold model of binary belief.

Does this show that the threshold view makes a place for significant deductive constraints on rational belief? It seems to me that it does not. For one thing, it is not clear why "n-wise consistency" principles should be intuitively attractive, from the point of view of describing ideal rationality. Of course, there is intuitive reason to impose the probabilistic constraints on graded belief upon which the limited-consistency principles supervene. But considered apart from the probabilistic constraints, there's nothing attractive about principles that one can believe inconsistent sets of beliefs only so long as they contain at least 17, or at least 117, members.

Moreover, when one moves to consider closure principles, the threshold model does not support similar limited versions of closure. As we've seen, one of the motivations for taking deductive constraints seriously is to account for intuitions such as the following:

> If an ideally rational agent believes both P and (P ⊃ Q), she believes Q.

Suppose we tried to advance a limited closure principle as follows: if Q is entailed by any *pair* of an ideally rational agent's beliefs, then the agent believes Q. This would seem to answer to the intuition above. But it would also amount to imposing an unlimited closure requirement. For any two beliefs will entail their conjunction; and, once that is admitted as a belief, it may in turn be conjoined with a third belief, etc., until the agent is required to believe any proposition that is entailed by any finite number of her beliefs. This is, of course, incompatible with the threshold account of rational binary belief, as the lottery cases demonstrate.[10]

Thus it seems that, insofar as sub-certainty threshold accounts of binary belief are plausible, we should look not to deductive constraints, but to probabilistic constraints, if we are to find plausible formal conditions on rational belief. We've seen above that a similar

[10] Indeed, the burden of Kyburg's (1970) "Conjunctivitis" is to cast doubt on the Conjunction rule for rational belief—that if an agent rationally believes P and rationally believes Q she must also believe (P & Q).

lesson holds for certainty accounts of binary belief. We've also seen that it is not plausible to unify belief by identifying graded beliefs with particular binary beliefs. Summing up, then, it seems that, while no unified account of belief is fully compelling, to the extent that graded and binary belief could be unified, the formal constraints that characterize ideally rational belief would likely be probabilistic.

Still, given that even the threshold account considered above is intuitively problematic, it is worth seeing whether a view of binary belief that made it more independent of graded belief could provide a home for deductive logical constraints. Such a view would, of course, divorce the two kinds of belief in a fundamental way. But several writers have advocated just this sort of divorce.

2.3 Bifurcation Accounts

Bifurcation accounts hold that binary beliefs are different in kind from graded beliefs—that neither is a mere species or special case of the other. Such accounts may be urged for various reasons. For one thing, bifurcation may allow for a better fit with some aspects of our ordinary assertion and attribution practices. In lottery cases in particular, we are reluctant to assert unqualifiedly "This ticket will not win," even when the lottery is large. Those who would tie binary belief closely with unqualified assertion may take this as important evidence against identifying binary belief with high confidence.[11] And there are other cases—in particular, those of apparently rational scientists discussing fairly comprehensive

[11] Maher (1993, 134) and Kaplan (1996, 127) explicitly support their bifurcation accounts in this way. Others, however, see assertability as tied to knowledge rather than belief (see Unger 1975, ch. 6; Williamson 1996; and DeRose 1996). DeRose, for example, would attribute belief in lottery cases such as the one described, holding that unqualified assertions would be improper because "this ticket won't win" would violate a counterfactual tracking-style requirement for knowledge (i.e. you would have the same belief even if you were holding the winning ticket).

theories—when unqualified assertions seem to be made about claims in whose complete truth no one should have very high confidence, given the history of science.[12]

Moreover, it must be acknowledged that even ordinary belief-attributions seem strained in lottery-type situations. Suppose that we know that Yolanda holds a ticket in a lottery she knows to be large, and that she has no special information about her ticket. Suppose we also know Yolanda to be highly rational. We would not hesitate to attribute to Yolanda a high degree of belief in her ticket not winning. But we might hesitate to say, flatly, "Yolanda believes that her ticket won't win." And if we asked Yolanda herself "Do you believe your ticket's a loser?" it would seem at least somewhat unnatural for her simply to reply "Yes."[13]

If unqualified assertion is taken as a mark of belief, then our ordinary assertion practices also seem to fit uneasily with threshold accounts in a way that is independent of lottery-type cases. Often, our willingness to make unqualified assertions seems to depend on aspects of the context quite independent of the likelihood of the relevant proposition's truth. Suppose, for example, that ten minutes ago I chatted in my driveway with the neighbors who live on either side of my house, after which I saw them disappear into their respective houses. I know that neither had plans to leave soon, but I haven't been watching their driveways. Someone knocks on my door by mistake, wanting to speak to my left-hand neighbor about an upcoming concert. I might well say to the person, "Jocko's at home next door." On the other hand, when a doctor knocks on my door by mistake, wanting to consult my right-hand neighbor on an emergency life-and-death decision about her relative, I would not say "Yolanda's at home next door." I might say that she's probably at home, or even almost certain to be at home, but I wouldn't just say unqualifiedly that she was at home. Some have

[12] Maher (1993) argues along these lines; his views on theory acceptance will be discussed in Chapter 4.

[13] On the other side, though, as DeRose (in correspondence) points out, it would also be unnatural—maybe even more so—for her simply to reply "No."

used this sort of case to suggest that belief is sensitive to what is at stake in a given matter, and not just to the agent's degree of confidence that the proposition is true.[14]

From our perspective, however, the most interesting argument advanced in support of bifurcation accounts is not about fit with ordinary assertion and attribution practices. It is a more theoretical one, which applies directly only to rational (or reasonable, or warranted, or justified) binary beliefs. If the standard deductive consistency and closure constraints apply to rational binary belief, then it cannot be rational to believe that a given large lottery will have a winning ticket, while simultaneously believing of each ticket that it will not win. Now no one seems to want to deny that it can be rational to believe that a big fair lottery will have a winning ticket. But various philosophers have devised conditions on justification, warrant, acceptability, etc., that are expressly intended to preclude rationally believing of any particular ticket that it will lose, no matter how high the odds. If we reject the requirement that rational belief be absolutely certain, it is argued, then only a bifurcated account can possibly allow for binary beliefs to be made subject to rational constraints of deductive consistency and closure. Thus bifurcation views are endorsed precisely because they allow for rational binary beliefs to be governed by logic.[15]

Now since the deductive constraints apply only to *rational* beliefs, it might be doubted that their application could be used to argue convincingly for a conclusion about the metaphysics of binary belief in general. And some epistemologists who have defended deductive constraints in the face of lottery examples do not seem to have had metaphysical conclusions explicitly in mind. BonJour, for example, holds that in lottery cases one does not have a *fully justified belief* that one's ticket will lose. He points out that a belief's degree of justification cannot then be correlated with the

[14] See e.g. Nozick (1993, p. 96 ff.).

[15] For examples of arguments against sub-certainty threshold views of rational belief, see Kaplan (1996, 93 ff.), Maher (1993, 134), Pollock (1983), Lehrer (1974, 190–2), and BonJour (1985, 54–5).

probability of the belief's truth. But he does not explicitly address the question of whether binary belief itself—the sort of belief with which he is concerned—is an attitude that goes beyond having a certain degree of confidence in the relevant proposition.[16]

It is worth seeing, then, whether a unificationist about the metaphysics of belief—say, a sub-certainty threshold theorist—could accommodate the deductive constraints on rational belief. He would have to admit that, when an agent's degrees of belief in the members of the inconsistent set of lottery propositions are each over the threshold, the agent does indeed harbor inconsistent (binary) beliefs. However, he would hold that the beliefs in question were not fully rational (or completely justified, or warranted).

This line seems unpromising to me. Our unificationist must acknowledge that the agent contemplating the large lottery *should* have a high degree of belief in, e.g., the proposition that ticket no. 17 won't win. But if her having a high degree of belief in this proposition is fully rational, and if having the binary belief is nothing over and above having a high degree of belief, then it is surely something of a strain to suggest that the binary belief that ticket no. 17 won't win is not rational in this case. It is, after all, one and the same attitude toward one and the same proposition—that is the essence of the unification approach.

The threshold theorist might try to differentiate between different types of rationality: the agent's attitude might be claimed to be degree-rational but not binary-rational. Surely there is nothing wrong with acknowledging different dimensions of rationality, and admitting cases where they give different verdicts about the same object. For example, one might reasonably think that having a

[16] In BonJour's description of the belief component of knowledge, there is no obvious mention of any factor going beyond degree of confidence: "I must *confidently believe* ..., must accept the proposition in question without serious doubts or reservations. Subjective certainty is probably too strong a requirement, but the cognitive attitude in question must be considerably more than a casual opinion; I must be thoroughly convinced...." (1985, 4). I should note that this description is part of an account for which he claims only approximate correctness; nevertheless, the reservation he expresses about the belief component is unrelated to the present issue.

certain religious belief, or a belief in the fidelity of one's friend, was pragmatically rational, but that having exactly the same attitude toward exactly the same proposition was epistemically irrational.

Nevertheless, I think that this sort of move will not work in the present case. For in calling an agent's attitude toward a certain proposition irrational one is endorsing a perspective from which the agent's attitude toward that proposition is undesirable. In the present case, since binary-rationality is an epistemic notion, the perspective will have to be an epistemic one. But it is clear that there is nothing at all to be said, from any epistemic perspective, against our agent's high degree of confidence in the proposition that ticket no. 17 will lose. There is no epistemic perspective from which her having a lower degree of confidence would be at all preferable. Thus it turns out that a unifying view cannot accommodate deductive constraints on binary belief by distinguishing degree-rationality from binary-rationality: doing so would deprive binary-rationality of all normative force.

It seems, then, that the plausibility of imposing deductive constraints on rational binary beliefs does have implications for the metaphysics of binary belief in general. Unless we hold binary belief equivalent to certainty, the imposition of the deductive rational constraints requires that binary belief be divorced from graded belief in a fundamental way. Believing a proposition must involve taking some attitude toward it that is wholly distinct from one's confidence that the proposition is true.

Of course, the power of any argument that sought to support a bifurcated metaphysics of belief in this way would depend directly on showing independently that it was plausible to impose the deductive constraints in the first place. Whether this can be done is a question that will be examined closely in the following two chapters. At this point, we can say that a bifurcated metaphysics of belief may find some support in our ordinary assertion and attribution practices, and is a prerequisite to the imposition of the standard deductive constraints on rational belief.

The questions of how, and whether, rational belief is constrained by logic are intimately connected with the question of what belief is. On either a graded or a binary conception, logical relations among propositions can be used to constrain rational belief. But the two conceptions invite quite different ways of doing so: the binary conception invites the imposition of deductive closure and consistency, while the graded conception invites the imposition of probabilistic coherence.

Both conceptions of belief have at least prima facie claims to describing important features of our epistemic lives. But the relation between the two kinds of belief is not obvious. Unifying the two conceptions by seeing one kind of belief as a special case or species of the other seems plausible only in one direction (assimilating binary to graded belief). This would leave probabilistic coherence as the fundamental formal constraint on rational belief. In fact, the more plausible route to unification, the sub-certainty threshold approach, is incompatible with taking full-blooded deductive constraints as normative requirements on rational belief. It seems, then, that imposing the deductive constraints requires adopting a fundamentally bifurcated view of belief; the next two chapters will explore this possibility. Probabilistic constraints, on the other hand, may find a home on either a unified or a bifurcated metaphysics of belief; the plausibility of probabilistic constraints will be explored in subsequent chapters.

3 DEDUCTIVE CONSTRAINTS: PROBLEM CASES, POSSIBLE SOLUTIONS

3.1 Intuitive Counterexamples

DEDUCTIVE consistency and deductive closure provide attractive constraints on ideally rational belief (for convenience, I'll combine these conditions under the heading "deductive cogency," or sometimes just "cogency"). The constraints of deductive cogency require, as we've seen, quite a specific conception of belief: a binary, yes-or-no attitude, which must consist in something over and above the agent's having a certain degree of confidence in the truth of the believed proposition. Presumably, if these constraints play an important role in epistemology, this role will be illuminated by an understanding of what the point of binary belief is. But before examining questions about the purpose or significance of this sort of belief, I'd like to look at some cases that directly challenge the legitimacy of taking rational belief to be subject to demands for deductive cogency. I think that the lessons these cases teach us prove useful in examining the question of whether the point of binary belief can motivate a cogency requirement.

Let us begin with a classic case often referred to as posing the "Preface Paradox."[1] We are to suppose that an apparently rational person has written a long non-fiction book—say, on history. The body of the book, as is typical, contains a large number of assertions. The author is highly confident in each of these assertions; moreover,

[1] A version of this argument was first advanced by Makinson (1965).

she has no hesitation in making them unqualifiedly, and would describe herself (and be described by others) as believing each of the book's many claims. But she knows enough about the difficulties of historical scholarship to realize that it is almost inevitable that at least a few of the claims she makes in the book are mistaken. She modestly acknowledges this in her preface, by saying that she believes that the book will be found to contain some errors, and she graciously invites those who discover the errors to set her straight.

The problem for deductive consistency is obvious. We naturally attribute to our author the belief, apparently expressed quite plainly in the preface, that the body of her book contains at least one error. We also naturally attribute to her beliefs in each of the propositions she asserts in the body of the book. Every one of these beliefs seems eminently rational. Yet the set of beliefs we have attributed to her is inconsistent. Moreover, the fact that our author, apparently quite reasonably, fails to believe that the body of her book is entirely error-free puts her in violation of the closure requirement.[2]

The problem here is clearly related to that posed by the lottery cases. There, if the agent believes of each ticket that it will lose, then he is precluded by consistency from believing that the lottery will have a winning ticket, and is required by closure to believe that it won't. But in at least one important way, the intuitive challenge posed by the preface case is sharper. In lottery cases, as we have seen, people do have some reluctance to assert flatly of their ticket that it will lose, and perhaps even to acknowledge believing that it will lose; this gives some encouragement to those who would deny belief—or rational belief—in these cases. But this is certainly not true of the individual claims made in the body of our author's book. Thus the dominant cogency-preserving response to preface cases

[2] In stating the preface case initially, I have been careful to be explicit about the fact that the belief expressed in the preface applies only to beliefs expressed in the body of the book, i.e. not to beliefs expressed in the preface itself. This is to avoid introducing complications of self-reference. In what follows, I will sometimes omit "the body of" for the sake of readability; I hope the intention is clear.

does not involve denying that the author rationally believes each of the claims in the body of the book.[3]

Defenders of cogency have thus typically wanted to deny that the author is rational in believing what I'll call the "Modest Preface Proposition":

> *Modest Preface Proposition.* Errors will be found in the body of this book.

Denying rational belief in the Modest Preface Proposition clearly does not have the initial plausibility of denying rational belief in lottery-case propositions of the form "ticket *n* won't win." Admittedly, there would be something odd about a preface that baldly

[3] An exception is Sharon Ryan's treatment of preface cases (1991), which argues that in all but certain very unusual cases, books by hard-working, intellectually responsible authors always contain unjustified claims. Ryan acknowledges that if one writes a short and simple book on addition for first graders one might succeed in writing a book with only justified claims in it; but in that case, of course, it does not seem intuitively that the modest preface statement would be rational. I doubt that this line can succeed in solving the preface problem for rational belief; it would seem to depend on setting the standards for rational belief excessively high. Given that responsible scholarship can easily produce rational beliefs about history (and not just about, e.g., elementary arithmetic), there is no barrier to producing history books consisting of rationally believed propositions. And given that rational belief need not be infallibly produced, a substantial book of such propositions may easily be highly likely to contain errors. One might object that, if we stick to Ryan's terminology of justified (rather than rational) belief, and interpret justification strongly, it is plausible that actual historians do typically make claims in their books that are not justified. Suppose this were granted. It still seems that cases posing the preface problem can be constructed easily. We might substitute for an academic historian a more humble sort of researcher: one who looks up telephone numbers, say, for a political campaign. I take it as uncontroversial that carefully looking up a person's phone number in the directory counts as a method of acquiring a justified belief as to what the person's phone number is. Now suppose that our campaign worker is incredibly scrupulous: she uses a ruler to line up the names and numbers, and she looks up each number on two separate occasions before entering it on her campaign list. It seems to me that, with respect to any particular number we might choose from her campaign list, she is justified in believing that it is correct. (If it is objected that memory limitations will preclude her from harboring hundreds of beliefs such as "Kelly Welly's number is 555–1717," we may concentrate on her beliefs such as "The 317th phone number on my list is correct.") Of course, phone directories are not infallible. Thus we need only make the campaign list long enough, and it will be overwhelmingly likely that there will be a mistake in it. And it seems that if our campaign worker understands this, she cannot be, as Ryan would have to claim, justified in believing her list to be error-free.

stated "This book contains errors!" But it does not seem at all odd to write in a preface, "In time, errors will be found in this book, and when they are, I hope that they will quickly be brought to light." And if an author is asked, "Do you believe that any errors will be found in your book?" or even, directly, "Will any errors be found in your book?" there is nothing at all unnatural about her saying, simply, "Yes."[4]

One might think that the problem could be avoided by taking a fairly lenient view of the demands made by formal constraints on rational belief. Taking deductive cogency as a rational ideal need not commit one to calling irrational anyone who falls short of the mark. One might, for instance, take the import of the cogency constraint to be something like this: If an agent can easily recognize that her beliefs are not cogent, and it is also clear how her beliefs could be revised to restore cogency, then rationality requires restoring cogency.[5]

In the preface case, however, the inconsistency is blatant. So is the lack of closure, insofar as we make the obvious supposition that our author lacks belief in what I'll call the Immodest Preface Proposition:

> *Immodest Preface Proposition.* The body of this book is 100% error-free.

Moreover, it is clear that consistency can be restored simply by the author's dropping the belief that her book will be found to contain errors. And closure—insofar as the stipulated facts of the case go—could be accomplished by the author's adopting a belief in her own book's historical inerrancy. Thus, the violation of these constraints does not seem to be excusable, even on a moderate reading of the force of the constraints.

In addition, irrespective of how easily the departures from cogency could be discovered or repaired, the preface case does not

[4] A quick check of the prefaces in books lying around my reading chair revealed the following sentences following immediately after an author's listing of those to whom he is philosophically indebted: "Their stimulus is largely responsible for what may be of interest in this book. The mistakes are all mine" (Mellor 1971, ix).

[5] Maher advocates this sort of position on consistency (1993, 134–5).

seem to be the kind of case in which, even though there are certain improvements possible in the agent's beliefs, those improvements might rationally be forgone. The changes in the agent's beliefs here that would restore cogency do not strike us as possible improvements at all—they are as intuitively irrational as they are easy to formulate. Thus it seems that the preface case provides a strong prima facie argument against taking deductive cogency as a rational ideal, on any reading of how violations of formal constraints relate to rationality.

Finally, one might try to dissolve the difficulty that preface cases present by distinguishing carefully between first- and second-order beliefs. One might insist, for example, that closure would not require any second-order belief about the first-order beliefs expressed in the book. It might be conceded that closure would require belief in the conjunction of the book's first-order claims, but that belief might be held to be distinct, from the logical point of view, from the Immodest Preface Proposition.[6]

This line seems unpromising to me. For it seems clear that an author who knew what she had said in the body of her book could realize that this conjunction was materially equivalent to the second-order claim of inerrancy for the body of the book. Once she has accepted the equivalence, closure will take her from the conjunction to the second-order claim.[7]

One might try to block this line of reasoning by taking the normative force of formal principles of rationality to be conditioned

[6] Simon Evnine (1999) uses something like this strategy in attempting to undermine a version of the preface problem framed as a challenge to the principle that rational beliefs are closed under conjunction.

[7] Evnine concentrates on an extended version of the preface problem in which an agent reflects not just on the beliefs in a given book, but on all of her beliefs—"a vague, ill-defined and exceedingly large" set (1999, 206). This makes less transparent the relations between the first- and second-order beliefs. Evnine uses this to argue that lottery cases provide a stronger challenge to the closure-under-conjunction principle he ultimately seeks to defend. But given that book-oriented preface cases avoid the cited difficulty, and that lottery cases have their own difficulties (as noted above), it seems to me that the challenge presented by preface cases is stronger (indeed, Evnine's solution to the lottery problem does not apply to preface cases).

by the agent's limitations, so that closure would require that the agent believe propositions entailed by her other (rational) beliefs *only if the entailed propositions could be entertained by the agent*, in the sense that the agent could bring the content of the entailed proposition clearly before her mind. One might then argue that our author might well not be able to entertain the massive conjunction of all the claims in the body of her book, and thus that even if closure is taken as a rational ideal, our author is not required to believe in her book's inerrancy.

This objection also seems unpersuasive. It is undoubtedly true that ordinary humans cannot entertain book-length conjunctions. But surely, agents who do not share this fairly superficial limitation are easily conceived. And it seems just as wrong to say of such agents that they are rationally required to believe in the inerrancy of the books they write. Clearly, the reason that we think it would be wrong to require this sort of belief in ordinary humans has nothing to do with our limited capacity to entertain long conjunctions.

Moreover, even if we restrict the closure principle to entertainable propositions, restrict our attention to ordinary agents, and distinguish scrupulously between first- and second-order beliefs, the preface problem can be developed. Surely an ordinary author who was paying attention could entertain the conjunction of the first two claims in her book, and recognize the material equivalence of this conjunction and the claim

(1) The first two claims in my book are true.

She would then be led by closure to believe (1). She could then easily entertain the conjunction of (1) and the third claim in her book. Our limited closure principle would then dictate believing that conjunction. Recognizing the equivalence of this believed conjunction with the claim

(2) The first three claims in my book are true

would lead, by similar reasoning, to belief in (2), and so on, until the belief in her book's inerrancy is reached. It must be granted that only an agent hard-up for entertainment would embark on such a process. But it is certainly not beyond normal cognitive capabilities, and the inerrancy belief seems no less irrational for having been arrived at by such a laborious route.

3.2 Consistency without Closure?

Suppose it is granted that in preface cases it would be irrational for the author to believe that the body of her book is 100% error-free. Assuming that the author might yet be fully rational in believing each of the claims she makes in the body of her book, this would seem to require giving up closure. Still, it might be thought that ordinary ways of thinking and talking make the preface-based case against consistency somewhat weaker.

Although most authors would be highly reluctant to assert the inerrancy of their books—and not just out of false modesty—it is also true that many authors would be reluctant to assert "This book contains errors." This might be taken as showing that authors typically lack belief in the Modest Preface Proposition. (It is, I think, unarguably natural to say "This book undoubtedly contains some errors." But it might be claimed that "undoubtedly" signals that the agent is expressing a degree of confidence rather than binary belief. And it might be claimed that what explains some authors' reluctance to make the former statement is precisely that unqualified assertions express binary beliefs, and these authors lack the relevant belief.)

Of course, even if we accept the claim that the reluctant authors lack the relevant belief, this would not show that they were *rational* in withholding belief. But a position that mandated withholding belief in these cases might seem easier to swallow than one that required authors to have a positive belief in the inerrancy of their own (current) scholarship. Might we salvage a partial defense of deductive constraints in the face of preface cases by retreating to the position that consistency is a rational requirement, even if closure is not?

This strategy avoids some of the implausible consequences of requiring full deductive cogency, but it seems to me that more than enough implausibility remains to undercut the value of the retreat. To see this, let us fill out a bit more fully the case of one particular "moderately immodest author"—one who does not assert (or believe) the Modest Preface Proposition, but who also does not assert (or believe) the Immodest Preface Proposition. This will allow us to see more clearly what the constraint of consistency by itself mandates in preface cases.

Let us suppose that Professor X, our moderately immodest author, sees himself as a solid historian. He would never write something in a book that he didn't believe, or something for which he didn't have very good evidence. But he also sees himself as a bit less neurotic than certain of his colleagues, in the following way: unlike them, he is free of a perverse fetish for endless minute and typically redundant fact-checking. He knows that each of his previous books has contained some minor errors of detail; this has, of course, allowed certain critics to exercise their nit-picking skills. But this does not bother Professor X much. After all, his books have been influential, and the broad conclusions they have reached are, he believes, entirely correct. Moreover, Professor X would point out that *every* book in the field—even those written by certain persnickety colleagues—has contained at least a few minor errors of detail. Indeed, he believes that writing a completely error-free book in his field is virtually impossible.

Given this background, it is not surprising that whenever a new book comes out—even a book written by a scholar he believes to be more meticulous than he himself is—Professor X believes that the new book will be found to contain errors. Time and time again, these beliefs have been borne out. And now, suppose that Professor X is studying a catalogue, in which his forthcoming book is being advertised alongside the new offerings from Professors Y and Z (both of whom have taken unseemly pleasure in pointing out niggling little mistakes in Professor X's previous works). He shows the catalogue to a nearby graduate student, chuckling, "I can't wait until someone finds all the little mistakes in Y's book."

"You believe Y's book has mistakes in it?"

"Of course I do. Why wouldn't it?"

"Do you believe that mistakes will be found in Z's book as well?"

"Yes! And I must admit, I'm looking forward to it. These anal-retentive types get so upset when they're caught in the most trivial errors! Look—all my books have had some minor errors in them. But you see, that's virtually inevitable, and it's no big deal. I'm not as careful as Y or Z, but my reputation is—well, I'm sure you see what I mean..."

"So—your new book here—do you believe that it has any little errors in it?"

"No."

Perhaps not everyone will share my intuitions here, but I think that Professor X's last statement would strike most people as an obvious joke. And the reason for this is that to take this statement at face value would be to attribute gross irrationality to him. Given the comparisons our author willingly makes between his work and that of Professors Y and Z, it is simply not rational for him to believe their books to contain errors, but not to believe the same about his own book. The fact that withholding belief seems so clearly irrational here—that rationality would seem to require Professor X to believe the Modest Preface Proposition—provides powerful evidence that deductive consistency is not a rational requirement.

One might object that the last line in the above dialogue—a simple "no"—may be taken to indicate actual disbelief, rather than the weaker suspension of belief. It is true that "I don't believe P" does often express belief in P's negation, and not just an absence of positive belief that P. Thus, the defender of imposing consistency but not closure might claim that the appearance of irrationality here is due to the impression that our author has adopted a belief in the Immodest Preface Proposition—a belief that is not required by consistency.

I think that this doubt can be dispelled by making the last line in the dialogue a bit more explicit. Suppose that, in answer to the student's asking if he believed that his new book contained any errors, Professor X had replied:

"No. I don't believe that my own book contains errors. I don't believe that it's error-free either. I'm just up in the air on that one."

It seems to me that this line is less funny only because it's pedantically drawn out. The attitude expressed by a literal reading of the dialogue is still absurd: our author believes quite firmly that every book in the field published so far, including his own, has contained multiple errors; he believes on these general grounds that Y's and Z's new books contain errors; he readily acknowledges that his own new book was written less carefully than Y's and Z's; and yet—somehow, unaccountably—when it comes to his own new book's inerrancy, he has no belief, one way or the other. This failure of his to draw the same conclusion about his own book that he so willingly draws about Y's and Z's books, when the evidence for the conclusion about his book seems to differ only in being somewhat stronger, strikes me as virtually a paradigm case of irrationality. Thus, I do not think that a defender of deductive consistency can escape the intuitive problem illustrated in the dialogue by noting that "I don't believe P" often means "I believe not-P."

Another objection to taking the dialogue as providing a serious intuitive challenge to consistency might be that its main character is too cavalier about getting things right to be ideally rational. Thus the fact that it seems wrong to impose consistency on *his* beliefs does not undermine the claim that consistency is required for ideal rationality.

I think that this objection misses the mark in two ways. First, it is not at all clear that our author's degree of caution in forming historical beliefs falls short of the rational ideal. It does clearly fall below a Cartesian standard whose achievement would preclude even the slightest possibility of error. But the Cartesian standard is not the appropriate rational standard for historical beliefs. So, while it is clear that Professor X is not the most epistemically cautious person in his field, this does not show that his level of caution is sub-ideal.

Second, a similar situation could be constructed with a person who is at the epistemically cautious extreme in the field—say, Professor Y—as its central character. Professor Y might be more troubled by the inevitability of minor errors in history books. But she would presumably share certain key beliefs with her less cautious colleague: that all previous books in the field, including her own, had contained errors; and that all the new books by other very careful writers—Professor Z, for example—will be found to contain errors. If she refused to draw a parallel conclusion about her own new book, while acknowledging that she had no evidence that her current scholarship was more careful than her past scholarship (or that of Professor Z, etc.), then it would seem to me that this refusal to treat such epistemically similar cases on a par was clearly irrational.

Two final points should be mentioned in evaluating the strategy of responding to the preface problem by giving up closure and trying to save only consistency. First, in giving up on closure, one would lose a major part of the motivation cited by some defenders of imposing deductive constraints on binary belief. Pollock, for instance, takes his fundamental epistemological assumption to show that the epistemic importance of arguments requires a closure principle.[8] Thus some defenders of deductive constraints would find the envisioned retreat unsatisfactory, even if it did avoid some sharply counterintuitive consequences.

Second, the motivation behind the retreat flowed from the intuitive strangeness of saying, flatly, "This book contains errors." But as

[8] See Pollock (1983, 247 ff.); Kaplan (1996) argues for a similar point. This argument will be examined in detail in the next chapter.

mentioned above, it is not clear that unqualified assertions express belief, rather than claims to knowledge. Thus one might well want to explain some of the awkwardness of the flat assertion as flowing not from lack of belief, but from failure of a condition on knowledge that goes beyond rational belief.[9]

In sum, then, it seems to me that, while there may be something to be said for a position that imposes deductive consistency but not closure as a condition on rational belief, retreating to such a position does not help much to reduce the severity of the preface problem.[10]

3.3 Extent and Severity of the Intuitive Problem

It seems that the best response to the preface problem, if one wants to impose deductive constraints on belief, will involve biting the whole bullet: holding not only that our author should refrain from believing that his book contains errors, but that he should positively believe his book to be entirely error-free. Indeed, this is the tack taken by supporters of deductive constraints on binary belief such as Pollock, Maher, and Kaplan.[11] Before moving on to discuss how

[9] See Williamson (1996, esp. sect. 3, and also 2000, ch. 11) for arguments that assertion is tied to knowledge rather than belief—even reasonable belief. DeRose's (1996) explanation of failure of assertability in lottery cases would seem to apply to preface cases at least as easily. On this view, our modest author realizes that if her book were, luckily, error-free she would have all the same evidence for her own fallibility, and thus would still believe that it contained errors. Thus she would judge herself not to *know* that her book contained errors, and would be unwilling to assert unqualifiedly that it did.

[10] Another kind of retreat that would allow for intuitively rational beliefs in preface cases would be to impose only a limited consistency constraint. We saw that Kyburg (1970) showed that on a sub-certainty threshold view of belief, inconsistent sets of beliefs in lottery cases cannot be smaller than a certain size (where the size depends on the chosen threshold). Hawthorne and Bovens (1999, 241–64) make a similar point about preface cases. But as noted above, this sort of limited constraint is essentially just an artefact of the probabilistic constraints on degrees of belief; it does not provide an interesting independent principle for rational binary belief.

[11] In a recent treatment of preface cases, Adler (2002, ch. 7) explicitly claims only that the Modest Preface Proposition is not believed. But if I understand his position correctly, it would also sanction belief in the Immodest Preface Proposition.

one might either explain away the unintuitiveness or show that it must, in reflective equilibrium, be accepted, let us examine exactly what kind of unintuitive consequences the imposition of deductive cogency requires.

Consider a variant on the case examined above. Suppose that Professor X is a "fully immodest" author, who respects not only deductive consistency but deductive closure in the preface case. We'll join the dialogue part way through, after Professor X has expressed his firm beliefs that (1) every previous book in the field (including his own) has contained multiple errors; (2) he's not as careful a scholar as Y or Z; and (3) the new books by Professors Y and Z will be found to contain errors. Let's start at the point when the graduate student poses the crucial question:

"So—your new book here—do you believe that it has any little errors in it?"

"No. I believe that this book of mine is completely error-free."

"Wow! Is that a first?"

"Absolutely. I believe that mine is the first book ever in the field that is 100% devoid of falsities."

"Is this because your subject this time was particularly amenable to accurate scholarship?"

"Not at all."

"Were you especially careful this time?"

"Certainly not. I'll leave the obsessing over trivia to Y and Z."

"But doesn't all this make it pretty likely that there are at least some little mistakes?"

"Of course; it's overwhelmingly likely that my book contains many errors."

"But you just said you believed..."

"Right. I believe that my book does not contain even one little error."

I think that most people would be incredulous at Professor X's claims, if they took them as something other than a deliberate joke at his own expense. Taken literally, he attributes to himself a set of beliefs that are, to my mind at least, patently irrational. The intuitive irrationality shows itself in at least two different ways.

(a) Unequal Treatment

The first way in which our author's beliefs seem intuitively irrational is a somewhat stronger version of the problem noted above in the first version of the example. Professor X comes to the conclusion, based solidly on excellent evidence, that errors will be found in other scholars' books. Simultaneously, he comes to exactly the opposite conclusion about his own book. Yet his evidence for errors in his own book seems even stronger than his evidence for errors in the books of others. Intuitively, this strikes us as irrationally treating similar cases differently. Given that the unequal treatment seems explicitly designed to favor the agent's epistemic assessment of his own beliefs, part of the intuitive irrationality here seems to spring from something akin to epistemic arrogance. But even if arrogance is not the agent's motivation, the unequal treatment seems indefensible.

One might object here that it isn't quite right to say that Professor X has better evidence for the existence of errors in his book than for errors in Y's and Z's books. After all, in the case of his own book, he already believes the claims it advances. Thus, these beliefs—which do entail the inerrancy of the book—give him a reason to think his own book to be error-free. And he clearly lacks a parallel reason for thinking the same of Y's and Z's books.

We should be careful not to allow this objection to sound stronger than it is. The objector cannot be claiming that Professor X has some reason for thinking that his own book is *less likely* to contain errors than are Y's and Z's books. In fact, as stipulated, our author is, quite rationally, more confident that his own book will contain errors than that Y's or Z's book will. So the unequal treatment is not a matter of our author's having any privileged reason for rational confidence in the truth of his book's claims.

To see this point clearly, consider what happens in cases where Professor X reads an authoritative book in a field somewhat distant from his own. In many such cases, he straightforwardly believes what the book says. Let us consider such a case, in which Professor X's formation of new beliefs is a typical case of rationally accepting claims on authority. His acceptance of the book's authority does not, of course, mean that he thinks it to be an infallible source of truth. But it seems obvious that claims made in a book by a respected authority may meet the standards for rational belief.

Now, before reading the book, Professor X will, as before, have the reasonable belief that errors will be found in it. And after reading the whole book, his assessment of the probability of errors being found need not change at all. The book may not have an especially sloppy or especially careful style, and Professor X may have no special information that confirms or disconfirms the book's claims. Yet, the moment he reads (and believes) the final claim in the book, deductive cogency will require Professor X to execute an abrupt epistemic about-face, abandoning his original belief about the existence of errors in the book, and adopting instead the contrary belief that his colleague's book is 100% error-free!

The point here is not just to give another instance where deductive cogency demands intuitively irrational belief. The point is to make even clearer that the unequal treatment we saw in the original example cannot be justified by citing Professor X's special evidence for the beliefs in his own book. The reasons Professor X has for believing the claims he reads in this new book are no better than his reasons, before reading, for believing that the claims in the book were true. It's really just the bare fact that he now has adopted the beliefs expressed in the book that grounds his newly generous epistemic assessment of it.

Of course, there is an asymmetry between a case in which Professor X has read a new book, and one in which he hasn't yet read it. In the former case, once he has believed the book's claims, he does believe propositions that entail that the book is error-free. But that asymmetry cannot solve the intuitive problem facing the advocate of deductive cogency. For this asymmetry does not motivate differential beliefs about the existence of errors in the two books in any way that is independent of the basic demand for deductive cogency. Perhaps there are independent arguments for deductive cogency that are strong enough to override its counterintuitive consequences. But whether or not this is true (an issue that will be examined in detail in the next chapter), it seems clear that the unequal treatment demanded by deductive cogency presents an acute intuitive difficulty.

(b) Internal Incongruity

The second way in which Professor X's beliefs in the example beginning the section exhibit intuitive irrationality is in the incongruity (reminiscent of Moore's Paradox) of "It's overwhelmingly likely that my book does contain many errors, but I believe that it doesn't contain even one."[12] Of course, it may well be that (as in Moore's cases) the sentence in question is not itself strictly inconsistent. But it is also clear that it is not a sentence one would expect to hear from any ordinary person—even a frankly arrogant scholar. In fact, it is hard to imagine anyone saying such a thing in ordinary life, at least with a straight face. (One can imagine it being said in jest: someone who has planned a 3 pm picnic and has just seen the morning weather forecast might say, "I know it's almost certain to keep raining all day, but I choose to believe that the sun will come out by 3 pm!" The humor here derives directly from the irrationality of the self-ascribed belief. If the person really did believe, in the face of all the evidence, that the sun would come out by 3 pm, we would not hesitate to deem her belief irrational.)

Moreover, not all of the incongruities are clear Moore-style examples mixing first-person belief-ascriptions with direct claims about the world. For Professor X's beliefs presumably will include both "My book very likely contains errors" and "My book does not

[12] A milder incongruity, even closer to Moore's Paradox, arises just from the imposition of consistency: "It's overwhelmingly likely that my book does contain many errors, but I don't believe that it does." Kaplan notes the Moore Paradox flavor of cases like the ones under discussion. His defense of the rationality of such beliefs will be examined in the next chapter.

contain errors." Like the more Moorean example, this does not quite constitute a contradiction; nevertheless, it certainly is not an intuitively rational combination of beliefs, and a person making both assertions categorically would strike anyone as bizarre.[13]

Thus it turns out that, in preface cases at least, imposing deductive cogency on rational belief conflicts quite dramatically with our ordinary practice. Ordinary rational people do not in such cases make the categorical assertions, or self-ascribe the beliefs, that deductive cogency would require. Nor would we be at all inclined to suspect that such beliefs were had by paradigmatically rational friends. In fact, the thought that the cogency-mandated beliefs are, or would be, rational in such cases is intuitively quite absurd. The bullet that must be bitten here is clearly substantial.[14]

3.4 Extent and Severity, cont.: Downstream Beliefs and Everyday Situations

Of course, if intuitive counterexamples to a general theory are few in number and peculiar in structure, we may have less reason to worry. Perhaps BonJour had this sort of point in mind when he consigned the preface problem to part of one footnote in a substantial book defending a coherence theory of justification which

[13] This actually may understate the problem. As we will see in the next chapter, it is not clear how the two claims in question can rationally be believed by a cogent agent without giving rise to an explicit contradiction.

[14] Some advocates of cogency are clearly concerned about this sort of intuitive problem, but not all. In Pollock's detailed treatments of preface cases (1986; 1990, ch. 9), he takes the problem to be just that of showing that the mechanism of "collective defeat"—by which he denies warranted belief in lottery cases—does not generalize in a way that would deny an author's warranted beliefs in the individual propositions asserted in her book. Pollock accomplishes this (in a way that stems from the fact that the lottery propositions are mutually negatively relevant, unlike the propositions asserted in a history book), and takes it to provide "a satisfactory resolution to the paradox of the preface" (1990, 253). Pollock does not seem to count it as part of the paradox that his account would bestow warrant on, e.g., our author's belief that his own book is the first 100% error-free contribution to the field.

includes deductive consistency as a necessary condition on coherence. After mentioning a different objection to imposing consistency, he writes: "And there are also worries such as the Preface Paradox. But while I think there might be something to be said for such views, the issues they raise are too complicated and remote to be entered into here" (1985, 240, fn. 7).

Yet it seems to me that the magnitude of the problem posed by preface cases should not be underestimated. One aspect of these cases that is not typically emphasized enough is that adopting a belief in the Immodest Preface Proposition is not something that occurs in an epistemic vacuum—especially for an agent who is deductively cogent. Let us think in concrete detail about Professor X's beliefs, to get a feel for some of the possible "downstream" effects of his believing the Immodest Preface Proposition.

In the situation envisaged, it would certainly be rational for Professor X to have the general belief that writing a completely error-free book would require being extremely careful and meticulous (which he knows he is not), or being amazingly lucky. Thus if Professor X's beliefs are to be deductively cogent, he must believe:

(a) I am amazingly lucky.

And believing himself amazingly lucky is only the tip of the iceberg. Given his belief that writing a completely error-free book would be such an unprecedented achievement, Professor X must take this fact into account in forming his beliefs about the future. For example, given his information about his colleagues, he is undoubtedly rational in believing that if no errors can be found in his book, Professors Y and Z will be in for a big surprise. In fact, he may well be rational in believing that if anyone wrote a completely error-free book in his field, it would soon lead to adulatory reviews, prestigious speaking engagements, and opportunities for professional advancement. He would then be committed to believing the following predictions:

(b) Professors Y and Z are in for a big surprise.

(c) I will soon receive adulatory reviews, prestigious speaking engagements, and opportunities for professional advancement.

Indeed, in many different situations, the Immodest Preface Proposition will combine with background beliefs to result in all manner of strange beliefs. Suppose, for example, we add to our story that, several decades ago, the Society for Historical Exactitude established a medal and a monetary prize, to be awarded to the first book in the field advancing substantial new theses in which no errors had been discovered one year after its publication. Although by now the monetary prize has grown to substantial proportions, the award has gone unclaimed, for reasons that Professor X understands only too well. Clearly, the fact that this award exists does nothing to weaken the requirement imposed by deductive cogency that Professor X believe the Immodest Preface Proposition. Now Professor X knows that his new book, like all of his works, advances substantial new theses. And since he believes that his new book is also the first such book to be 100% error-free, he should presumably believe:

(d) One year from now, I will be able to wear the SHE medal and start spending the SHE prize money.[15]

And this monetary belief will have consequences of its own. Given that Professor X knows himself to have particular automotive aspirations, he might well be justified in believing that if he comes into a sufficiently tidy sum, he'll buy an Alfa-Romeo. In that case, he should presumably now believe:

(e) In one year, I'll be driving an Alfa.

Clearly, it would be easy to pile on intuitively irrational beliefs in Professor X's case as long as one wanted to. And it's also clear

[15] If the Society for Historical Exactitude seems too fanciful, it is worth noting that the connection between the feat of writing an error-free book in Professor X's field and the consequence of receiving large sums of money might easily be forged by more common mechanisms, such as job offers at well endowed universities. Thus, one could tell substantially the same story involving

(d') One year from now, I'll be working at a prestigious university and start spending my generous salary.

that it would require little ingenuity to come up with countless other cascades of intuitively irrational beliefs in different cases in which people obey deductive cogency by believing Immodest Preface Propositions. The structure of the problem involves cogency twice: first, cogency requires belief in an incredibly improbable proposition; then it requires belief in whatever propositions are entailed by conjoining the improbable proposition with ordinary reasonable background beliefs. The result is a chain reaction of cogency-mandated beliefs that are if anything more clearly irrational than the Immodest Preface Proposition itself.

Now it must be admitted that this plethora of intuitive irrationalities is almost certain to be short-lived. When the first reviews of his book appear, Professor X will come to realize that his book is not, after all, error-free; beliefs such as (a)–(e), which were spawned by the Immodest Preface Proposition, will then vanish as well. But, far from alleviating the intuitive strain that the example provides, this point should serve to underline an additional dimension of wackiness involved in holding beliefs such as (a)–(e). For a whole nest of beliefs like this will spring up anew each time Professor X publishes a book. Given deductive cogency, Professor X's knowing that similar sets of beliefs have arisen with the publication of each of his books, only to be leveled by the book's first reviews, will not serve in the slightest to diminish his epistemic duty to embrace each subsequent set of sanguine predictions.

Clearly, the problem posed by believing the Immodest Preface Proposition is not merely that this one belief is itself intuitively irrational. Adoption of this belief will have a strong tendency to spread ripples of intuitive irrationality throughout various parts of a deductively cogent agent's belief system. For the affected agent, the epistemic difficulties are quite severe.

It might be thought, though, that the malady, however grave, is at least a rare one. Might the defender of cogency take comfort in the claim that situations posing preface-like difficulties are rare, or remote in some way from ordinary epistemic life?

It must be admitted that actual prefaces worded in the way the classic example requires are fairly uncommon. But it is not at all uncommon, or remote from ordinary epistemic life, for people to write books expressing their beliefs. And many of these books are written in fields such as history or biography, where the number of factual details involved in a book makes it quite likely that the book contains errors. Whether or not these books have prefaces, deductive cogency would require all of their authors to believe them to be 100% error-free. Of course, very few of these authors have any such belief. And of those who have formed some belief on the question, the great majority undoubtedly believe that their books will be found to contain at least some minor errors. Thus, it turns out that there are quite a few real people in ordinary situations who have preface-style beliefs about their books—beliefs that, while intuitively quite rational, directly and obviously violate deductive cogency.

Moreover, beliefs of this structure are not restricted to authors of books in detail-oriented fields. Many of those who have reflected even briefly on their own fallibility believe that at least one of their (other) beliefs is mistaken.[16] Some would undoubtedly hedge on the issue, saying only that they probably had at least one mistaken belief. But I suspect that only a tiny minority believe—as closure would dictate—that every single one of their beliefs on every topic is correct. Most would, I think, share Henry Kyburg's sentiment:

[16] Again, the restriction to their "other" beliefs is intended to avoid self-reference. Evnine (1999) claims that this restriction, while needed to avoid self-referential paradox, poses a problem of its own. He claims that

*Something I believe, other than this belief, is false

"makes invidious distinctions among our beliefs and gives a special status to some that it does not give to others, namely, exemption from possible error" (p. 205). This objection should not, I think, be persuasive. After all, given that (as Evnine supposes) our reason for thinking ourselves fallible is empirical, it is not surprising that we have more reason to doubt some sorts of beliefs than others. People's beliefs about their hair color or addresses, for instance, are much less prone to error than their beliefs about, e.g., details in history. We in fact have excellent (empirical) reason to believe, about anyone, that she has at least one false first-order belief, and thus we have excellent (empirical) reason to believe that anyone who accepts the limited expression of epistemic modesty expressed by * is correct in so doing. So exempting * from its own scope is not, as Evnine claims, a case of "special pleading."

"I simply don't believe that everything I believe on good grounds is true, and I think it would be irrational for me to believe that" (Kyburg 1970, 59).[17]

Finally, apparently rational violations of deductive cogency may crop up even in quite ordinary situations that do not involve second-order beliefs. When I go to bed at night, I believe that the newspaper will be on my front porch by 6:30 the next morning. I don't, of course, have absolute certainty about this matter, but I've been taking the paper for years, and have more than enough experience of its reliability to make this a reasonable belief. I also have just as good reason to believe that the paper will be on my porch by 6:30 two mornings hence, and equally good reason to believe that it will be there three mornings hence, etc. If you ask me, "Do you believe that the paper will be on your porch by 6:30 am seventeen days from today?" I will answer affirmatively, without hesitation. I think it quite plausible to attribute to me, for each $n < 366$ at least, the belief that the paper will be on my porch by 6:30 am n days from now. But I also know that, on rare occasions, the paper does not arrive in the morning. Thus I also believe that, on some morning in the next year, the paper will fail to be there by 6:30. Clearly, these intuitively rational beliefs violate consistency.[18]

[17] See also Klein (1985, 131); Kitcher (1992, 85); Foley (1993, 165); Nozick (1993, 78). Kyburg suggests that those who are tempted to deny the natural view here are misled by quantifier confusion: "of everything I believe, it is correct to say that I believe it to be true; but it is not correct to say that I believe everything I believe to be true." It is interesting to think about the downstream effects of following cogency rather than Kyburg here. If the implications of Professor X's belief about having written an error-free history book seem wild, they are surely tame compared with the consequences flowing from an agent's belief that she—and presumably she alone, among all the people who have ever lived—is correct about every single matter on which she has a belief.

[18] This example is loosely based on an example in Hawthorne and Bovens (1999, 242). It might be objected that I do not really have the belief about, e.g., day 17 until I am asked, and thus, that if I haven't separately considered each day I do not have all the beliefs claimed for me in the example. Of course, we would not want to insist that for me to have a belief, I must be actively entertaining it. So the objection would have to be that I don't even have dispositional beliefs here. This objection seems weak to me, for two reasons. First, I think that we do typically attribute beliefs to people in propositions that they have not actually entertained, but which they would unhesitatingly agree with if asked. For example, I think that we would attribute to most people the belief that

Moreover, if I were to adopt the closure-mandated belief that, in this coming year, the paper would never once fail to be there by 6:30 am, my belief would be intuitively irrational.

The newspaper case may be thought to resemble lottery cases more than it does preface cases. But it is, I think, worth developing independently of the standard lottery case, for the following reason: in the standard lottery case, as we have seen, there is some intuitive reluctance to assert flatly "My ticket won't win," or to self-attribute the associated belief. In the newspaper case, my telling a house guest "The paper will be on the porch by 6:30 tomorrow" is entirely natural, as is my self-attribution of the correlated belief. Thus, our ordinary intuitive judgments about particular beliefs in the newspaper case seem to me to provide a clearer objection to deductive cogency than do our intuitions in the classic lottery case.

In sum, then, it seems that the intuitive challenge posed by apparently rational beliefs in preface-like and lottery-like situations is a strong one, in two dimensions. The beliefs demanded by cogency in some of these situations are not just slightly suspicious intuitively; they strike us as wildly irrational. And the situations in which intuitive rationality and deductive cogency conflict occur all the time, for ordinary agents in ordinary epistemic circumstances. Clearly, the intuitive burden imposed by deductive cogency cannot easily be shrugged off.

3.5 Undermining the Counterexamples?

Of course, even if it is acknowledged that our pre-theoretic judgments in the troublesome examples are firm ones, and that the counterintuitive cases are neither rare nor *recherché*, the existence of

there are more than 17 stars in the sky—even though not very many of them have had occurrent beliefs in this particular proposition. Second, I certainly could consider each of the day-specific newspaper propositions that figure in the example, and come to believe each, by any reasonable standard.

a large class of strikingly counterintuitive examples set in ordinary epistemic life does not in itself settle the issue. For one thing, it can always be argued that the reasons for imposing deductive cogency are so strong that our intuitive judgments about these examples should be overridden in reflective equilibrium. Assessing this sort of argument will depend on assessing the general arguments for imposing cogency; I'll turn to that in the next chapter. But a defender of deductive cogency might try a different tack. She might try to *undermine* (rather than override) our intuitive judgments in the apparent counterexamples, by showing, *on cogency-independent grounds*, why, e.g., it would be rational for Professor X to believe his book to be 100% error-free, or why it would be irrational for him to believe that some mistakes will be found in his book, or why I should not believe that the paper will be on my front porch tomorrow (or, perhaps, why I should believe that, this year at last, the paper delivery will never fail). Most such attempts I have seen focus on lottery examples; this is natural enough, since we do have some intuitive reluctance to claim belief of each ticket that it will not win. Below, I'll consider two such attempts, and will then examine one that focuses on preface cases.[19]

(a) Guilt by Association

One idea that has struck several authors as attractive in dealing with lottery cases focuses on the fact that such cases involve a set of beliefs (*a*) which, given what else is known (or rationally believed), contains at least one false member, and (*b*) whose member beliefs are very similarly based. The set of beliefs of the form "ticket *n* will lose" is of this sort. The idea is that, when the support that one has for each of a set of propositions does not significantly distinguish

[19] There may be no very deep difference between arguments that seek to undermine our intuitive judgments in the troublesome cases on cogency-independent grounds, and those that seek to override our intuitive judgments in a way that depends on seeing cogency as essential to rational belief. I am separating them here mainly for expository convenience.

among them, and one believes that at least one of these propositions is false, then that support is insufficient for rational belief in any one of them.[20] Let's call this the Guilt by Association (GBA) principle.

Clearly, this principle will not help out with preface cases; it would be generally inapplicable (since the beliefs in the body of the book will not generally be supported in indistinguishable ways), and it would not yield the desired result anyway (we do not want to deny that authors can rationally believe the claims made in their books). So the GBA principle cannot be a complete answer to the intuitive problems with deductive cogency.[21] Nevertheless, it does seem to have the advantage of meshing with our reluctance to attribute beliefs in lottery tickets' losing, and thus has some claim to providing non-question-begging motivation for preserving cogency in one important range of cases.

It is crucial to note that the principle is not a bare assertion of a no-known-inconsistency requirement; it does not come into play whenever one has a set of beliefs such that one knows that one of them is false. The beliefs in question have to be relevantly similar, so that, as BonJour (1985, 236) puts it, the agent has "no relevant way of distinguishing" among the beliefs in the set.[22]

[20] See BonJour (1985, 235–6); Ryan (1996); and Evnine (1999). Ryan's version is a bit stronger, in that it forbids justified belief in the similarly supported propositions even in cases where one is not justified in believing that at least one of these propositions is false, but where one is justified in withholding belief about whether at least one of them is false. I should also note that, while Evnine is directly addressing rational belief per se, BonJour and Ryan mean to be giving conditions on justified belief in the sense, roughly, of "meeting the justification condition of knowledge." I don't want to enter into the issue of whether this degree of justification corresponds to rational belief. Since I'm concentrating on the conditions for rational belief rather than knowledge, I'll just examine whether this basic approach can solve the problem that lottery cases pose for consistency constraints on rational belief.

[21] It might generalize a bit beyond standard lottery cases. For example, it might be thought to help with the newspaper case, if one thought that the right reaction to that case involved denying that I am rational in believing that the paper will be on the front porch tomorrow.

[22] A similar requirement is explicit in Evnine's statement (1999, 207). No such requirement is explicitly made in Ryan's official statement of her principle, but it seems implicit in her response to certain examples, and in her justificatory remarks on her principle (Ryan 1996, 130–5). Nelkin (2000) shows convincingly that Ryan's principle would be implausible without this requirement.

Now one worry one might have is expressed by Dana Nelkin (2000, 385): that principles of this sort are "so finely tailored to lottery-like cases that they are limited in their ability to explain what is really responsible for our lack of knowledge or rational belief in those cases." Without some deeper motivation, the GBA approach might be dismissed by opponents of consistency requirements as an ad hoc response to lottery examples.

One might press the intuitive motivation question further by asking why it should be relevant that one's grounds for the beliefs in question be very similar. It can't be simply because the agent knows that the grounds can fail; after all, our whole problem arises only within a context in which we're assuming that rational belief does not require infallible grounds. The thought must instead be something like this: "If an agent knows that a certain set of propositions contains a false member, she cannot rationally believe all of them. But since the grounds for believing the propositions are so similar, she has no non-arbitrary way of picking one not to believe. Thus, she cannot rationally believe any of them." But if this is the motivation for the GBA strategy, it will not help at all to undermine the intuitive counterexample to deductive cogency. For it presupposes that it cannot be rational to believe a set of propositions when one knows that one of them is false. And this would seem to beg the question in favor of imposing deductive consistency.

Moreover, non-standard lottery cases reveal that the GBA approach does not even succeed at the limited task of squaring all lottery cases with consistency requirements. Consider a lottery in which different tickets have different (but always small) probabilities of winning. In such a lottery, there will be relevant differences among the propositions in the falsity-containing set. Thus the GBA principle will not apply, at least in a straightforward way.

Ryan (1996, 132–3) does consider an example of this general sort. In a million-ticket lottery where one knows that "the fifty blue tickets sold have a much higher probability of winning than all the rest," Ryan holds that one must suspend judgment about the blue tickets, but one may believe of the others that they won't win.[23] But whether or not this is an intuitively reasonable thing to say in Ryan's case, other unequal-probability cases will be harder for the GBA approach to handle. One might, after all, have a one-guaranteed-winner lottery in which each ticket had a different (but small) chance of winning. Here there is no set of competing beliefs that are epistemically indistinguishable. Thus the GBA principle would seem not to apply in this sort of lottery case at all.

If this is right, then it is even harder to see why we should think that the GBA approach provides a plausible way of defusing lottery-based counterexamples to deductive cogency. Principles that deny rational belief in lottery propositions may, as we've seen, derive some independent support from meshing with what we are intuitively inclined to say about lottery tickets. But it now seems unlikely that the GBA approach provides the correct explanation for, e.g., our reluctance to say "I believe my ticket will lose." After all, we would be no less reluctant to say this sort of thing in an uneven-probability lottery case where the GBA principle is inapplicable. It seems, then, that those who want to undermine our intuitive judgments in the counterexamples to deductive cogency should look elsewhere.

(b) Banning Purely Statistical Support

Another lottery-inspired approach holds that a proposition may not be believed rationally if one's grounds for belief are, in some sense, purely statistical. Of course, any consistency-saving approach to

[23] Ryan's thought here, I take it, is that it is highly probable that some blue ticket will win. Thus, if one believed of each blue ticket that it would lose, one would have a set of relevantly similar beliefs which were such that one had good reason to believe that one of them was false. One cannot say the same of the white tickets, since it is not highly probable that one of them will win. It's not clear to me that this way of handling the case fits with Ryan's general objectives. She wants justified belief to serve as a sufficient condition (given truth and the absence of Gettier conditions) for knowledge. The present case does not seem like a Gettier case. Yet if the holder of a white ticket claims to know that her ticket will lose, and we find out that another white ticket wins, it does not seem that the knowledge claim was correct.

the lottery cases will have the consequence that high probability is not sufficient for rational belief. But some such accounts would, e.g., allow statistical support to suffice for rational belief in the absence of defeaters. The present idea is that statistical support is for some reason incapable of making belief in a proposition rational, even absent any special circumstances that might compromise that support in some way. Some writers (e.g. Cohen 1988, 106 ff.) have rejected statistically based beliefs as candidates for knowledge, and some have urged the same for rational belief. Here I'll concentrate on Nelkin's recent defense of the idea that statistical support is insufficient for rational belief.[24]

One advantage of this sort of position over GBA is that it seems less ad hoc, in that it applies straightforwardly beyond the standard lottery cases. For example, it applies unproblematically to the unequal-probability lottery considered above. It also does not seem to beg the question; after all, it applies to cases that pose no threat to consistency requirements, such as lotteries in which there will probably be no winner. In many cases where we have statistical reasons for thinking a certain event to be highly improbable, we do seem reluctant to make flat assertions (or self-attribute beliefs) to the effect that the event will not occur. Insofar as this reluctance can be taken to show lack of rational belief, we have some independent motivation for the approach of banning purely statistical support (BPSS).

Of course, those who take assertability as tied to knowledge rather than belief will find this last motivation questionable.[25] And there is something at least curious in the basic BPSS idea. After all, no one thinks that statistical support is irrelevant to rational belief, and everyone acknowledges that it comes in degrees. Then why, one

[24] Nelkin (2000) clearly separates the belief and knowledge cases, and advocates related solutions for both. A related proposal for rational belief is made in Kaplan (1996, 125 ff.).

[25] DeRose (1996) notes the lack of assertability in lottery-like cases, but argues (1) that this is due to failure of apparent knowledge rather than failure of belief, and (2) that the failure of apparent knowledge is not due to the fact that support is statistical, but to violations of a counterfactual-based condition.

might wonder, can't it be sufficient support for rational belief in some cases?

Nelkin offers motivation for BPSS that goes beyond preserving deductive constraints or meshing with assertion practices. When an agent believes rationally, he can "see a causal or explanatory connection between his belief and the fact that makes it true" (2000, 396). Nelkin would not require the agent to give a detailed description of the causal/explanatory connection. But the agent must be able to posit the existence of such a connection: "the key idea is that ... I can take myself to believe something *because* it is true" (2000, 398). This seems to rule out rational belief that one's lottery ticket will lose, for example, because it is clear that the ticket's losing would not explain or cause one's belief.

Now one initial worry is suggested by the requirement that an agent "believe something *because* it is true." If Nelkin required a rational agent to believe that the fact that would make her belief true actually *caused* her belief, then many obviously rational beliefs would be deemed irrational. So, for example, having turned the flame on under my skillet three minutes ago, I now believe—without touching or otherwise examining the skillet further—that it is hot. But the skillet's hotness does not cause my belief (nor does it seem correct to say that I now believe the skillet to be hot because it is hot). My belief is causally or explanatorily connected with the fact that would make it true, but not in the simple sense of the fact causing or explaining my belief. In countless cases of rational belief, our belief that P is caused or explained by factors which, in turn, cause or explain the fact that P.[26]

Let us interpret Nelkin's suggestion, then, to allow this indirect sort of causal/explanatory connection. On this interpretation,

[26] DeRose uses an example with a similar structure to reject Harman's claim that knowledge of P is made possible by an inference to the best explanation from one's evidence to P. DeRose points out that I can know, on the basis of reading in my copy of the paper that the Bulls won, that my neighbor's copy of the paper reports the Bulls' winning. DeRose points out that his own subjunctive account fares better here: if the neighbor's copy did not say that the Bulls won, then neither would mine, and so I would not believe as I actually do.

however, it is not clear that the suggested motivation for BPSS will succeed. Consider the agent's belief that his ticket will lose the lottery. This belief is explained by the agent's knowledge of the set-up and workings of the lottery in question. And it seems that this set-up was causally responsible for the lottery's outcome, including his ticket's losing. One might even take the fact that the lottery was set up in this way to provide an explanation of the fact that the agent's ticket did lose. After all, we do take the fact that a car is constructed in a certain way to explain the fact that it started when the ignition key was turned this morning (even though cars built this way do very occasionally fail to start). Of course, the issues surrounding the role of statistical connections in causation and explanation are extremely complex. But at this point, we have seen little reason to suppose that a causal/explanatory requirement will correctly weed out cases of statistically supported beliefs in a way that will help to motivate BPSS.

Moreover, it seems to me that BPSS faces a more acute problem that is independent of the motivational issue. Nelkin considers an example (from Harman 1986) in which Bill knows that Mary intends to be in New York tomorrow, and concludes from this that she will be in New York. But he also knows that Mary holds a lottery ticket with a one-in-a-million chance of winning, and that if her ticket wins, she'll go to Trenton instead. Harman's puzzle is about knowledge: Bill seems intuitively to know that Mary will be in New York, but not that Mary's ticket will lose. But the case provides, if anything, an even sharper challenge to Nelkin's view on rational belief. Intuitively, it seems quite rational for Bill to believe that Mary will be in New York. It also seems rational for Bill to believe that Mary will go to Trenton in the event that her ticket wins. Yet the BPSS approach does not allow us to say that Bill is rational to believe that Mary's ticket will lose, since that belief would have purely statistical support. Thus, the BPSS proponent is faced with the choice between embracing counterintuitive judgments or abandoning the deductive cogency requirement for rational belief.

Nelkin chooses to say that Bill lacks knowledge that Mary will be in New York. And although she discusses the case in terms of knowledge only, she would presumably deny that Bill's belief that Mary will be in New York is a rational one. But she hopes to mitigate the intuitive problem this sort of case poses by citing the rarity of such situations: "[I]t is important to note that we are not in Bill's situation very often. This means that it remains open that we often know where people will be (and not just where they are likely to be)" (Nelkin 2000, 407–8). Nelkin would presumably say the same about our rational beliefs about where people will be.

It seems to me, though, that denial of rational belief required here to maintain deductive cogency and BPSS will not be so easy to contain. It is true, of course, that we typically do not know that our friends are holding lottery tickets with the potential to derail their plans. But we do typically know, when a friend is driving to New York, that she'll be there only if she is not hit head-on by a drunk driver. Yet our grounds for believing that our friend won't be hit by a drunk driver seem to be of the purely statistical variety: we know that such events do sometimes occur, and we have no special reason to discount the possibility on our friend's particular route; however, we also know that accidents of this sort are incredibly infrequent. Moreover, it seems that in countless other cases our intuitively rational beliefs have this structure: we believe that P is true; we believe P is true only if Q is; and our reasons for believing that Q is true are merely statistical. We believe that we'll be at work on time; but we know we'll be late if (as occasionally but unpredictably happens) the bus breaks down. We believe that our car is parked where we left it; but we know that if (as occasionally but unpredictably happens) it has been stolen, it is somewhere else (see Vogel 1990 for this and several more examples). We believe that the Bulls won, but we know that if (as occasionally but unpredictably happens) the paper transposed the relevant scores, the other team has won (see DeRose 1996, 578–9). As Vogel points out, "Much of what we believe about the world beyond our immediate environments could be made false by some chance event we haven't yet heard of" (1999, 166).[27]

It turns out, then, that the BPSS strategy encounters a severe problem over and above the intuitive dubiousness of the idea that statistical support is somehow incapable of justifying rational beliefs. The strategy is particularly ill suited to providing a defense of deductive cogency. For it seems that, once one bans believing on purely statistical grounds, imposing deductive closure on rational belief forces one to embrace widespread skepticism—skepticism that vastly outruns any initial intuitive reluctance we have to claim belief in lottery-type propositions.[28] Yet it was just this intuitive reluctance that underlay the hope that BPSS could provide a way of independently motivating a cogency-respecting interpretation of lottery cases. Thus I think that the BPSS strategy cannot, after all, help to undermine these intuitive counterexamples to cogency.

(c) Sorites, Commitment, and the Preface

Preface cases present a harder problem for those who would undermine intuitive counterexamples to cogency. As we saw above, we do not attribute to ourselves the beliefs (and non-beliefs) required by cogency in preface cases; in fact, these beliefs (and non-beliefs) strike us as paradigmatically irrational. Thus, when faced with preface examples, defenders of cogency tend to argue at the

[27] Such examples are direct prima facie counterexamples to a closure principle on knowledge, since we are inclined to say that, e.g., we know who the President is, but don't know that he didn't have a fatal heart attack in the last five minutes. Vogel argues that this sort of example does not actually threaten closure; he argues that plausible ways of denying knowledge of the entailed proposition apply independently to the entailing proposition, so knowledge of the entailing proposition is not additionally threatened by denying knowledge of the statistically supported entailed proposition and then applying closure. For present purposes, there is no need to settle the issue of whether these examples involve some reason for disallowing rational belief other than that which would be provided by banning purely statistical support. If we have no special reason for disallowing beliefs supported on a "merely statistical" basis, then BPSS does not even get off the ground.

[28] See Vogel (1990), in which he terms this sort of threat "semi-skepticism."

general level that cogency must be a norm for rational belief, rather than trying to undermine our intuitive judgments in these specific cases. However, Simon Evnine (2001) provides cogency-independent arguments designed to show that it would be irrational for someone to believe that one of his (other) beliefs was false. In discussing Evnine's arguments, I'll adapt them to the specific case of the preface.

Evnine's first argument aims to show that the belief expressed in the problematic preface cannot be part of a fully rational set of beliefs. Let P be the Modest Preface Proposition: at least one of the claims in the body of this book is false; and let $C_1 - C_n$ be the claims in the body of the book. Suppose (for reductio) that the author's beliefs in P and all of the C_i are rational. If each of the C_i is rationally believed, there can be no particular one of them in virtue of which belief in P is rational—in other words, none of the C_i individually is such that the author is rational to believe *it* false. But if that is true, then, Evnine claims, P should still be rational if we excise one of the claims—say, C_n—from the book. If we accept this, then we can repeat the steps of the argument, excising a belief at a time, until we get to the point where the body of the book is reduced to C_1, and the author rationally believes P (which now applies, of course, to just C_1). But this is clearly absurd: it cannot be that both P and C_1 are rational; if the author is rational to believe C_1, he's not rational to believe P.

Evnine gives several versions of this basic argument, including some designed for countably infinite belief sets. In each case, the argument assumes an analogue of the premise used in the version above: that subtracting a rational belief from the body of the book cannot make the preface belief irrational (or, conversely, that adding a rational belief cannot make the preface belief rational). But this sort of premise is clearly reminiscent of, e.g., the claim that plucking a hair cannot make one bald. The problem with such assumptions with respect to rational belief (rather than baldness) might best be demonstrated by considering analogous cases having nothing to do with deductive cogency. Arguments may be constructed using essentially similar assumptions to derive radically skeptical conclusions about all sorts of intuitively rational beliefs.

Suppose, for example, that a murder takes place on a cruise ship. The ship's detective gathers all 317 people on board in the grand ballroom. The detective believes, on the basis of the extreme difficulty of anyone leaving the ship after the murder, that the murderer is in the room. It seems clear, on any non-skeptical view of rational belief, that the detective's belief may be rational. But the claim that this belief is rational would seem to be subject to a reductio exactly parallel to Evnine's reductio of the preface belief.

We may start by noting that there is no one person in virtue of whom the detective's belief is rational—i.e., there is no one particular person whom the detective rationally believes to be the murderer. So now, let us ask: can the detective simply excuse one of the 317 people from the ballroom, and remain rational in believing that the murderer is in the room? We may have some temptation to say "yes" to the first step here, if only because of the large number of suspects. But surely the general principle ("If the detective's belief is rational when *n* people are in the room, it will be rational when one of the *n* is excused") must be rejected. For it would allow the detective to excuse passengers one by one until she rationally believed, of the last remaining passenger, that he was the murderer. Detective work just isn't this easy.

In many ordinary cases, an agent has a belief that at least one of a very large set of objects has a certain property, and the agent holds this belief on grounds that are not specific to any of the members of the set. And in any such case, one can offer a sorites-style slide into skepticism. If we accept such offers, we will end up admitting that we cannot rationally believe, e.g., that someone on the ship committed the murder; that at least one student at the University of Vermont was born in March; that at least one book in the library has a chapter that begins on page 17; that we ate spaghetti on at least one day in 1998; etc. Clearly, if there is something rationally defective about the modest preface belief, it cannot be shown by this sort of argument.

Evnine also offers an independent way of explaining why we shouldn't believe the preface proposition. Unlike some other defenders of cogency, who admit that one must allow that it is probable that some of one's beliefs are false, Evnine denies even this. But then what about the inductive evidence provided by the beliefs of others, and one's own past beliefs? Evnine says that the inductive argument fails because one must be *committed* to one's own current beliefs, in a way that precludes thinking that some of them are false.[29]

Aside from questions about whether we should see beliefs as commitments of any sort, it is unclear why the sort of commitment involved in belief would (or could) undermine the rationality of using inductive evidence in the ordinary way to support the Modest Preface Proposition. We are, after all, quite willing to form beliefs on less-than-conclusive grounds. The sort of commitment that would block even the moderate degree of epistemic modesty involved in believing the Modest Preface Proposition would seem appropriate only if our standards for rational belief-formation were Cartesian. Moreover, it is hard to see what, on this view, our attitude should be toward propositions related to the Modest Preface Proposition. One should not, presumably, believe that, unlike everyone else (and unlike one's former self), one currently is employing special methods of belief-formation of a uniquely reliable sort. Should one, then, simply be supremely confident that one is now astoundingly lucky (though one would

[29] Evnine compares believing to promising. One may have broken some of one's promises in the past, but, he asks, "Can one now address the promisee and say that one is confident that one will fail to keep some of the promises one is currently making?" (Evnine 2001, 167). The analogy here seems strained to me. In the promise case, the commitment involved makes sense in large part because it is the agent himself who makes it the case that his promise is kept. In the belief case, the agent clearly cannot make it the case that her beliefs turn out to be true. For this reason, the breaker of even a sincere promise is typically morally culpable for reneging on his commitment. But the holder of a rational belief that turns out false has not thereby committed any epistemic sin.

appear to have no grounds for that assessment)? Or should one somehow refuse to form any opinion at all on how likely it is that one is possessed of special belief-forming methods or stunning epistemic luck? Nothing in the neighborhood of these thoughts seems even close to rational. And, more importantly, no such thoughts seem intuitively to flow from any sort of commitment one might undertake, merely in virtue of forming beliefs. Thus it seems to me that thinking about beliefs as involving commitments would do little to undermine our intuitive judgments in preface cases.

There are, no doubt, other ways of trying to undermine our intuitive judgments about rational belief in lottery and preface cases. But it is unlikely that defenders of cogency will succeed in removing the counterexamples' sting. Our pre-theoretic judgments, in the preface cases especially, are firm and stable. Thus, the best case for cogency will have to be made directly, and the consequent violence done to our pre-theoretic intuitions will have to be rationalized on the basis of the direct arguments for cogency. Let us, then, turn to examine those arguments.

4 ARGUMENTS FOR DEDUCTIVE COGENCY

It would be a mistake to dismiss deductive cogency merely on the basis of intuitive counterexamples, even if they are powerful and pervasive, and even if we see no way of undermining our intuitions in these cases. For it might turn out that anything we say on this topic will entail severe intuitive difficulties, and that rejecting cogency would carry an even greater cost than imposing it. After all, binary belief will, on any bifurcation account, be some propositional attitude whose point is not simply to reflect rational confidence in a proposition's truth. If we could be brought to see binary belief as an important and interesting component of epistemic rationality whose point requires deductive cogency, we might come to override our intuitions in the problematic cases.

In doing this, we might then seek some measure of reflective equilibrium by explaining the intuitions as resulting from a tendency to run binary belief together with another concept. For example, Mark Kaplan holds that our intuitive concept of belief is incoherent. We really have two separate notions: one of degree of confidence (or graded belief); and the other of acceptance (or binary belief). Graded belief alone figures in rational practical decisions. But binary belief has its own purposes, quite distinct from those of graded belief. And it is these purposes which lend importance to a variety of belief that is subject to the rational demand of deductive cogency.[1] Let us look, then, at some of

[1] Kaplan suggests that the Moore Paradox impression that one gets from assertions such as "I'm extremely confident that there are errors in my book, but I don't believe that there are any errors in it" stems from our confusing binary belief with a state of confidence. If belief in P doesn't require being confident that P is true, the sentence

the arguments that have been offered in support of deductive cogency.

4.1 ... and Nothing but the Truth

A first stab at explaining why binary beliefs should be consistent flows from the very core of the concept of epistemic rationality: that epistemic rationality aims at accurate representation of the world. A natural expression of this idea as applied to binary belief is that an epistemically rational agent seeks to believe what is true, and to avoid believing what is false. If an agent's beliefs are inconsistent, she is automatically precluded from fully achieving the second of these objectives. How can this be ideally rational?[2]

One problem with this argument is that avoiding false belief is not the only epistemic desideratum: if it were, it would be rational to reject all beliefs. Having true beliefs is also important. Moreover, we do not even want avoidance of false beliefs to take lexical precedence over having true ones: if it did, it would be rational to believe only those propositions of whose truth we were absolutely certain. Yet once one sees rationality as involving a balance between the two desiderata, the quick argument for cogency collapses. For it would seem inevitable—on any weighting of the desiderata against each other—that there will arise situations in which the best balance between the desiderata will be achieved only by failing to maximize with respect to either one of them. In preface cases in particular, refusing to adopt the Modest Preface Belief (that mistakes will be found in one's book) keeps alive the logical possibility that one will avoid all false belief. But it is incredibly unlikely that,

isn't problematic; see Kaplan (1996, ch. 4). Maher (1993, 153) expresses a similar view about the folk concept of belief.

[2] Lehrer (1974, 203) makes this argument, though he's since given it up; see his (1975). Foley (1987, 257–8) has a very nice critical discussion of this sort of argument, partly along lines similar to some of those offered below.

in rejecting the Modest Preface Belief, one will avoid a false belief. The overwhelmingly likely consequence is that one will have forgone a true belief, and thus achieved a poorer balance of truth over falsity. Insofar as having true beliefs is desirable, the Modest Preface Belief looks like an excellent candidate for adoption.[3]

Moreover, even the goal of avoiding falsity—on any natural interpretation—itself militates against treating preface cases as cogency would require. Achieving cogency in preface cases requires the adoption of the Immodest Preface Belief (that one's book is 100% error-free). Now adopting this belief does, of course, leave open the logical possibility of perfect error avoidance. In fact, adopting this belief could not spoil an agent's perfect record of error avoidance—if she had one. But on any natural interpretation of the goal of error avoidance, it does not reduce merely to valuing error-free belief sets above others. It distinguishes among the other, imperfect, belief sets, and values having fewer errors (and, perhaps, less important ones) over having more. Since the Immodest Preface Belief is almost certain to be false, the goal of avoiding error will itself tell against this belief's adoption.

Still, the thought that the pursuit of truth will in some way rationalize deductive cogency may seem attractive. A sophisticated version of this type of thought seems in part to motivate Kaplan's assertion-based account of binary belief, which is designed to support a cogency requirement:

> You count as believing P just if, were your sole aim to assert the truth (as it pertains to P), and your only options were to assert that P, assert that ~P, or make neither assertion, you would prefer to assert that P. (Kaplan 1996, 109)

Given the considerations rehearsed above, it is not clear how this analysis would favor imposing cogency. Suppose, for example, we asked what the rational author of a history book would

[3] Early examples of the basic decision-theoretic approach to binary belief are in Hempel (1960) and Levi (1967).

assert, with respect to the proposition that her book was error-free, if her sole aim were to assert the truth about this proposition. It would seem obvious—at first pass, anyway—that she should assert that her book was not error-free, since she is virtually certain that this proposition is true. But Kaplan does not intend the above-quoted passage to stand on its own; in particular, he wants to give a specific interpretation to the "aim to assert the truth":

> The truth is an error-free, comprehensive story of the world: for every hypothesis h, it either entails h or it entails ~h and it entails nothing false. This being so, the aim to assert the truth *tout court* is not one anyone can reasonably expect to achieve. But it is, nonetheless, an aim you can pursue—you can try to assert as comprehensive a part of that error-free story as you can. (Kaplan 1996, 111)

This interpretation of our epistemic goal—asserting as comprehensive a part of the error-free story as one can—draws our attention to the entire *body* of what one would be willing to assert, rather than to the individual propositions. Will this help us see how a cogency requirement could drop out of a desire to tell the whole truth and nothing but the truth? Would it license asserting a story containing the Immodest Preface statement, rather than an otherwise similar story containing the Modest Preface claim?

It seems to me that fixing our attention on whole stories does not in itself affect the argument significantly. True, refusing to assert the Modest Preface statement would leave open the bare logical possibility that the totality of one's assertions comprised a part of the error-free story. But of course, as Kaplan would certainly acknowledge, even this is not something anyone can reasonably expect to achieve. And if achieving total freedom from falsity is not a realistic option, it is hard to see what one would lose in foreclosing it. The total story that one ends up asserting is virtually certain to be a large part of the error-free story, plus a smaller budget of false claims. Including the Modest Preface Proposition in one's global story is virtually certain to increase the portion of the error-free story one asserts, whereas including the Immodest Preface Proposition is virtually certain to increase one's budget of false assertions.

The idea, then, cannot be just to come as close as possible to telling the entire error-free story—at least, not in the obvious sense of maximizing the truth and minimizing the falsity in the story one does tell: there is no direct road here which starts from the desire to tell a story as close to the global truth as possible, and ends with the strictures of cogency. The idea must be that there is some independent sort of value in telling (or believing) a cogent story *per se*. Other authors have expressed something like this idea. Van Fraassen (1995, 349) writes: "The point of having beliefs is to construct a *single* (though in general incomplete) picture of what things are like." Jonathan Roorda concurs:

> our beliefs are not just isolated sentences in a collection; they are meant to hang together, to tell a univocal story about the way the world is. It is this feature of belief which subjects it to the requirement of deductive cogency: we do not require the gambler to make sure that all of the propositions he bets on be logically consistent; but we do require of the storyteller that the logical consequences of what she has already said will not be contradicted as the story unfolds."[4] (Roorda 1997, 148–149)

It seems worth emphasizing that, insofar as this sort of consideration is to support deductive cogency in a way that goes beyond the advice to believe only what is absolutely certain, we seem to have left the desire for accuracy behind in a fairly dramatic way. For it's not only that the value of telling a cogent story fails to *follow from* the value of telling the truth. Cases such as the preface show that defending cogency would require that the value of telling a cogent story actually *trump*, or *override*, the value of veracity. In the next section, we'll look at what might be said for this sort of view.

[4] I should note each of these authors is defending an account of binary belief quite different from Kaplan's. On van Fraassen's account, belief entails certainty; on Roorda's, one believes only what one is certain of in at least some circumstances.

4.2 Keeping your Story Straight

Several writers have pointed out that we typically try to avoid asserting inconsistent sets of claims. An assertion-oriented account of belief, combined with the idea that rational belief is deductively cogent, would explain and justify this tendency. Now we've seen that the close ties between assertion and belief are contestable; assertions may represent knowledge claims rather than expressions of binary belief. But let's not pursue this issue here. Is there some special point in presenting to others (or even to one's self) a coherent picture of the world—a point that would outweigh the value of keeping the picture as accurate as possible?

It is important to keep in mind that our tendency toward maintaining deductive cogency in our assertions is far from absolute. Preface cases present dramatic examples in which our ordinary assertion practices violate cogency quite flagrantly. And if one ties assertion to belief, then our ordinary belief-attributing practices provide further violations of cogency; after all, everyday expressions of epistemic modesty such as "Everyone has some false beliefs, and I am not the sole exception" are made routinely, without upsetting ordinary believers/assertors in the slightest. Thus to use an assertion-based view to defend cogency, one would have to show that our ordinary beliefs (or assertion practices) in these cases are actually wrong for some reason. Is there something about the point of belief (or assertion) that makes our common-sense responses to these situations defective?

Kaplan, unlike many defenders of cogency, squarely addresses the intuitive challenge posed by preface cases. How, he asks, can the author of a history book be rational in asserting (and believing) that her book is 100% error-free, when she knows full well that this is extremely unlikely to be true?

> In outline, the answer is quite simple. Unless she wants to give up the writing of history, our author has no choice... [O]ne simply cannot assert an ambitious, contentful piece of narrative and/or theory *without* running

a very high risk of asserting something false. So our historian has a choice. She can decide that the risk is intolerable, in which case she will refrain from writing history. Or she can decide to tolerate the risk and pursue her profession. (Kaplan 1996, 118)[5]

Now the point here cannot be that there is some real-world obstacle faced by professional historians who fail to assert, or to believe, Immodest Preface statements. (In fact, it's not clear that professional historians ever make the sort of Immodest Preface assertion required by cogency.) As Kaplan would surely acknowledge, the profession of history tolerates Modest Preface assertions without batting an eye. Still, it is clearly true, as Kaplan points out here, that one cannot typically assert a whole ambitious theory without asserting something that's likely to be false. And this could be turned into an argument for something like the claims quoted above. If we could show that asserting (or believing) whole ambitious theories was required for doing history (or systematic inquiry in general) correctly, then we could support the claim that Immodest Preface assertions (beliefs) were in some way intellectually necessary.

Is assertion of (or belief in) entire, highly detailed accounts of the world a necessary part of inquiry? Some doubt is engendered by the fact that actual Modest Preface statements vastly outnumber Immodest ones. One might try to explain away appearances here. It might be claimed that people are really thinking, inside, "My book is the first one ever to be 100% correct, even though my claims are no less controversial, and I haven't used especially reliable methods, and I haven't checked my facts more thoroughly, [etc.] ..."—all the while disingenuously professing belief that the book will be found to contain errors, in order to appear modest. But

[5] I should note that this is not Kaplan's main argument for cogency. One might even interpret the quoted claim as simply presupposing cogency: if cogency were mandatory, then there would be a sense in which the historian would be required to believe the Immodest Preface claim. But I think that Kaplan intends to be saying something more here. He follows up the quoted passage by arguing against certain alternative attitudes that inquirers might take to ambitious theories, suggesting that the quoted claim is intended to do more than point out that Immodest Preface assertions are required, given the presupposition of cogency. Kaplan's main argument for requiring cogency will be discussed in the next section.

this suggestion strikes me as psychologically implausible in the extreme.

It might also be urged that scientists do sometimes make unqualified assertions of large theories, or, perhaps more commonly, describe themselves as believers in large and detailed theories. But should we take these claims as assertions that the theories in question will never be found inaccurate in even the tiniest detail? Consider theories about the origins of the first human inhabitants of the Americas. Such theories surely do rest on extremely large collections of detailed claims about diverse matters, including linguistic data on current inhabitants of America and adjacent continents, genetic information, physical measurements and chemical dating of particular fossilized human remains, analysis of stone tools and other artefacts found in certain locations, dating of fossilized bones from animals apparently butchered by stone tools, claims about climatic conditions and animal extinctions (themselves based on various sorts of archeological evidence), and more. Now suppose an archeological anthropologist says that she believes a theory according to which the first Americans came from Siberia over the Bering land bridge. Will this support the view that rational scientists believe that the sort of large comprehensive theories under consideration are completely true?

If we take the "Siberian origins" theory to be just the single claim that the first American came from Siberia, then our scientist's belief is not of the sort under discussion, since her assessment of the probability of such a single claim may well be quite high. Her profession of belief will support Kaplan's position only if her theory is taken to include a large number of detailed claims, such as those involved in the interpretations of countless specific bits of evidence of the sorts mentioned above. In other words, the theory must include enough so that our scientist—like Kaplan's historian—will rationally be extremely confident that it is not completely true. Now suppose we make our question clear by asking her explicitly, "Do you believe that in the entire theory—including such-and-such details about this linguistic item being causally connected with that one, this piece of rock being a tool fashioned at about such-and-such a time, the marks on this fossil being caused by scratches from a stone tool wielded by a human being, etc.—not one detail is incorrect?" Do we imagine that any scientist would answer affirmatively here? Or do we have any (pre-theoretic) intuition that it would be rational for her to have such a belief? If not, there seems little reason to think that rational scientists do really harbor the kind of beliefs that the argument in question sees as necessary for successful inquiry.

Moreover, in order to defend cogency as a rational requirement in the way envisaged, we would have to show more than that rational inquirers typically harbor beliefs to the effect that vast and richly detailed theories are completely true in every detail. It would have to be shown that this sort of belief plays a crucial role in inquiry. In other words, there would have to be some serious problems besetting inquirers who believed that even minor inaccuracies would ever be found in the details of their favorite ambitious theories. Scientists who merely believed that their favorite theories were *approximately* true in *most* respects would be at some sort of intellectual disadvantage. But it is hard to see what that disadvantage could be. Once a scientist or other inquirer has made all of the particular assertions involved in her ambitious theory, what is gained by her taking the extra step of asserting that her ambitious theory is absolutely flawless? It is hard to see what role such a performance would play in rational inquiry.

Of course, none of this is meant to deny that there is a scientific purpose in thinking and talking about big, detailed theories. There may well even be good reasons for scientists to form certain sorts of allegiances to such theories, to "work within" the systems that such theories provide. But all of this is quite compatible with believing that the theory one is developing is only approximately right: that the story it tells is largely correct; that the entities, processes, forces, events, etc., that it postulates are reasonably close to those that actually exist. In fact, it seems quite plausible to say that an important part of an inquirer's commitment to an ambitious theory is precisely to identify and correct those parts of the theory that are mistaken!

The claim that acceptance of whole theories plays a crucial part in science has been made in a different way by Maher.[6] He begins by endorsing Kuhn's observation that a highly successful large theory (paradigm) will not be rejected, even in the presence of anomalies, unless an alternative has been thought up. Maher notes that this may be explained by his account of acceptance: before development of the alternative, the anomalous evidence lowers the probability of the accepted theory. But since there is no alternative available, the fact that it offers a comprehensive account that's probably fairly close to the truth makes it rational to stick with it. When the alternative is dreamed up, there is a better option, and so the first theory is abandoned.

This seems entirely sensible, as far as it goes. But notice that the sort of commitment to theories invoked in this explanation need not include anything like the belief that the theory in question is true in every detail. The described commitment even seems compatible with the belief that the theory will be found to contain at least some inaccuracies. In fact, Maher notes that pre-Einsteinian physicists clinging to Newton's theory in the face of anomalies proposed *modifications* to Newton's theory, including modifications of the inverse-square law. This seems hardly the sort of behavior that would be expected of inquirers who believed Newton's theory correct in every detail, or even who refused to believe that it erred in any respect. Thus, while systematic inquiry may depend on investigators being guided by some sort of allegiance to a large theory, there is little reason to see this allegiance as incompatible with acknowledging the theory's imperfection.

[6] See Maher (1993, 169 ff.). Maher's notion of acceptance, like Kaplan's, is supposed to capture an aspect of our folk notion of belief (the other aspect being degree of confidence). Maher takes this notion to be subject to a deductive consistency requirement.

So far, then, we have seen no reason to think that either our intuitive reluctance to assert Immodest Preface statements, or even our willingness to assert Modest Preface statements, is misguided. Nothing we have seen so far about the role of big theories in inquiry seems to give a point to our asserting or believing massively conjunctive claims which we rationally regard as highly improbable. If a mandate for imposing cogency on binary belief is to be supported by some part of our intellectual practice, we will have to find it elsewhere.

4.3 The Argument Argument

A third strand of argument intended to support cogency focuses directly on how logical relations seem to inform rational belief through *arguments*. John Pollock writes:

> The main epistemological assumption from which everything else follows concerns the role of arguments in epistemological warrant. I have assumed that reasoning is a step-by-step process proceeding in terms of arguments and transmitting warrant from one step to the next in the argument. From this it follows that warrant is closed under deductive consequence ... (Pollock 1983, 247)

A "warranted" proposition, for Pollock, is one that an ideal reasoner would believe; he uses similar considerations to argue that warranted belief is deductively consistent. According to Pollock, arguments are as strong as their weakest links, and deductive inferences are completely warrant-preserving. Thus, a deductive argument from warranted premises must have a warranted conclusion.

Kaplan makes a similar claim about rational binary beliefs. He notes that, when a critic demonstrates via a *reductio* argument that the conjunction of an investigator's beliefs entails a contradiction,

> the critic thereby demonstrates a defect in the investigator's set of beliefs—a defect so serious that it cannot be repaired except by the

investigator's abandonment of at least one of the beliefs on which the *reductio* relies.

But if it is a matter of indifference whether your set of beliefs satisfies Deductive Cogency, it is hard to see how *reductios* can possibly swing this sort of weight. (Kaplan 1996, 96)

Kaplan applies a parallel point to constructive arguments:

[I]f satisfying Deductive Cogency is of no moment, ... the fact that we convince someone of the truth of each of the premises of a valid argument would seem to provide her no reason whatsoever to believe its conclusion. (Kaplan 1996, 97)

Pollock takes arguments as justificatory structures within an agent's cognitive system. Kaplan puts his points in terms of the interpersonal persuasive force of arguments, but it's clear that he sees such force, when it is legitimate, as flowing from a rational demand on each of us to have deductively cogent beliefs. Thus, for both writers, the challenge of accounting for the rational force of *arguments* should be understood as the challenge of accounting for the way in which rational belief seems to be conditioned synchronically by deductive logic.

To evaluate this challenge, we should ask whether the rational force that arguments actually have can be explained *without* invoking a cogency-governed notion of binary belief. I'd like to begin examining this question by looking at something that may at first seem beside the point. Let us see how deductive logic constrains rational *degrees* of belief, in situations where we see arguments as doing serious justificatory work. As we saw in Chapter 2, there is a natural way of constraining rational graded beliefs that flows directly from the logical structures of, and relations among, propositions: it is to subject rational graded beliefs to a norm of probabilistic coherence.[7]

[7] Of course, some would reject the notion of constraining graded beliefs in this way. For the present, I will assume (as do, e.g., Kaplan and Maher) that logical constraints on graded belief are legitimate—or, at least, that there is no in-principle objection to them that does not apply equally to cogency requirements on binary belief. In later chapters, I'll defend this assumption.

We can see right away that probabilistic coherence will force rational degrees of belief to respect *certain* deductive arguments: if P entails Q, then a rational agent's belief in Q must be at least as strong as her belief in P. Of course, this applies when P is a big conjunctive proposition. So if P is the conjunction of the premises of a valid deductive argument, and Q is its conclusion, then when a rational agent is very confident that the conjunction of the argument's premises P is true, she must believe the argument's conclusion Q at least as strongly. (If we think about this case in an interpersonal and diachronic way, we get a parallel result: my convincing someone to believe P strongly provides her with a reason to believe Q at least as strongly.)

The same mechanism works in *reductio* arguments. Consider a *reductio* aimed at rejecting Q, which is based on premises whose conjunction is P. In such a case, the conjunction P will be inconsistent with Q. Thus a rational agent's confidence in Q can be no higher than one-minus-her-confidence-in-P. So if she is confident to degree 9/10 in the conjunction of premises P, she must give no more than 1/10 degree of credence to Q. Similarly, her confidence in ~Q must be at least as high as her confidence in the conjunction of the *reductio's* premises. (Interpersonal-diachronically: if a critic points out to me that P is inconsistent with Q, and I am unwilling to give up my strong belief in P, I will have reason to give up my strong belief in Q.)

This sort of example shows that deductive arguments can have important effects on rational belief, even absent any cogency requirement—indeed, even absent any consideration of binary belief at all. So the Argument Argument does not come anywhere near showing that cogency requirements provide the only way for deductive arguments to gain epistemic purchase on us. But the examples above differ in two ways from what Pollock and Kaplan have in mind. First, they involve cases in which the agent not only finds each of the premises in an argument belief-worthy, but also finds the conjunction of the premises belief-worthy. Second, the belief-states described in the examples are graded, rather than binary. Does the phenomenon that is evident in these examples extend to cover the sorts of cases that seem to prompt the pro-cogency argument?

Let us focus first on multiple-premise arguments. Suppose that there is a long argument from P1–P*n* to C. In such a case, the fact that a rational agent believes each of the premises strongly does not necessarily give her any reason to believe C strongly. This is because having a high degree of confidence in the truth of each premise need not mean having a high degree of confidence in the conjunction of the premises. After all, the premises might be negatively relevant to one another, the truth of one making the truth of the others less likely (a simple example of this is seen in lottery cases, with claims of the form "ticket *n* won't win"). And even if the premises are independent of one another, their conjunction will typically be far less likely to be true than any one of them. For example, suppose that P1 is "the paper will be on my porch tomorrow morning," P2 is "the paper will be on my porch 2 mornings hence," and so on. If we take the set of such propositions up to P365, we get a valid argument for the conclusion "the paper will be on my porch every morning for the next year." But probabilistic coherence does not force a rational agent who strongly believes each of the individual premises considered separately to believe the conclusion at all strongly. For believing each of the premises—even strongly—does not rationalize strongly believing their conjunction.[8]

Thus, it is clear that the logical force of deductive arguments on graded belief does not obey the principle that Pollock endorses: it is not generally the case that arguments are as strong as their weakest single links. In cases where one is not certain of the premises of an argument, we get the following result instead: a deductive argu-

[8] The same point applies to *reductio* arguments. If {P1–P*n*, C} is an inconsistent set, this does not force one's rational degree of belief in C to be low, unless one is not only highly confident in each of P1–P*n*, but also confident in their conjunction. Since the issues below arise similarly for *reductios* and constructive arguments, I won't discuss *reductios* separately in what follows.

ment that depends on a great many uncertain premises will (ceteris paribus) be significantly less powerful than an argument that depends on fewer. But this does not strike me as clashing with our ordinary ways of thinking about arguments. Surely we feel less compelled by an argument with a huge number of uncertain premises than by an argument with only a few—even if no particular one of the premises in the huge argument is, considered by itself, more dubious than the premises in the short argument.[9]

Of course, none of this shows that the effect of deductive arguments on degrees of belief exhausts the legitimate epistemic role of these arguments. And I suspect that those sympathetic to the Argument Argument will feel that the above discussion sidesteps the main issue completely. After all, what's at issue is how deductive arguments affect rational *binary* belief, not how they affect rational

[9] Pollock and Cruz (1999) present an example designed to challenge probabilistic analyses of arguments. They consider an engineer designing a bridge: "She will combine a vast amount of information about material strength, weather conditions, maximum load, costs of various construction techniques, and so forth, to compute the size a particular girder must be. These various bits of information are, presumably, independent of one another, so if the engineer combines 100 pieces of information, each with a probability of .99, the conjunction of that information has a probability of ... approximately .366. According to the probabilist, she would be precluded from using all of this information simultaneously in an inference—but then it would be impossible to build bridges" (p. 107). Here Pollock and Cruz seem to be endorsing the engineer's simply relying on her conclusion about girder size to build the bridge—even though this conclusion is probably based on at least one false premise! At first, this might seem simply absurd, especially considering that the safety of the bridge depends on its girders having appropriate sizes. But of course, engineers do rely on the outputs of calculations with many inputs—inputs that are somewhat subject to error. Does this practice support something like Pollock's "weakest link" principle? It seems to me that it does not. Engineers in the sort of situation envisaged presumably believe that any errors in their calculational inputs are highly unlikely to be large enough to affect the end result significantly. If this is right, then the real belief relied upon in bridge-building is not some (probably false) belief in the correctness of an exact size specification that follows from the multiple measurements. It is rather the belief that any errors in the input values are small enough that using the calculated value is close enough to be safe. (If the engineer did not believe this—e.g. if she believed that any errors in her premises were likely to be large enough to have a significant effect on her girder-size conclusion—then, if the probability of such an error even remotely approached the 0.634 level that the example specifies, using the calculated value in building a bridge would be unconscionably negligent.)

degrees of confidence. The arguments we have with others, which we write about in books such as this one, or rehearse to ourselves when we take a critical perspective on our own beliefs, are not overtly probabilistic. So the fact that deductive arguments can affect rational degrees of confidence might seem quite beside the point.

This protest would be decisive if rational binary belief were completely insensitive to rational degrees of confidence. But we have no reason to suppose that this should be so—in fact, quite the opposite is clearly the case. And insofar as rational binary beliefs are informed by rational degrees of confidence, the effects that deductive arguments have on the latter may well have important consequences for the former. This is particularly obvious on a threshold view, according to which binary beliefs just are graded beliefs of a certain strength. If one's confidence in the premises of an argument puts one's graded belief in the conclusion above the relevant threshold, it will thereby have produced exactly the effect we are looking for.[10] Clearly, threshold views illustrate the possibility of deductive arguments affecting binary belief via their effects on graded belief.

Now threshold views are not the only binary belief model available; in fact, threshold models are often rejected explicitly by proponents of deductive cogency. Of course, in the present context it would beg the question to reject threshold accounts *because* they don't support cogency. But while cogency failure is surely the most common reason for rejecting threshold accounts of binary belief, these accounts may be criticized on cogency-independent grounds as well. For example, Kaplan (1996, 97–8) points out that if binary belief just is nothing more than a certain degree (call it n) of confidence, then it would be *impossible* (and not just irrational) to withhold belief in a proposition that one thought likely to degree greater than n. This would seem to render unintelligible (and not

[10] Foley (1993, 167 ff.) makes some related points about *reductios*. He points out that the effectiveness of a *reductio* directed against one of a set of claims depends on both the size of the set and the strength of support for, and interdependence among, its members. See also Weintraub (2001).

just bad) Descartes' advice not to believe what is merely probable. For given one's degree of confidence in a proposition, the question of whether one believed it would already have been settled.

Kaplan's sort of worry could be answered by a metaphysically sophisticated first cousin of the threshold view. One might take binary belief, as Descartes apparently did, to be accomplished by an act of the will—an internal assenting to a proposition.[11] This would allow for the possibility of willing assent, or failing to will assent, to propositions in ways that did not match up with any particular level of probability. But one might further hold—not altogether implausibly—that *rational* assent (and thereby rational binary belief) was governed by a threshold of rational graded belief. (Indeed, this seems to be the form of Descartes' suggestion, with the threshold for rational belief set at absolute certainty.) This sort of view allows for the metaphysical bifurcation of binary and graded belief, while allowing deductive arguments to affect rational binary belief via their effects on rational graded belief.

So if the point of the Argument Argument is supposed to be that deductive reasoning can in principle play no role in conditioning rational binary belief unless binary belief is subject to cogency, then the argument is simply wrong. Our rational responses to deductive arguments may seem on the surface to flow from a cogency requirement; perhaps this helps explain why many epistemologists have seen cogency requirements as so obvious as to need no defense. But this interpretation of the role of arguments, initially appealing though it may be, is not the only one available. For arguments affect the degree of confidence it is rational to have in a proposition's truth; and, on virtually any account, rational degrees of confidence can have important implications for the rationality of binary belief. Thus, there is another clear route by which arguments may gain purchase on our rational binary beliefs. And this route is completely independent of any requirement of deductive cogency.

[11] See e.g. the Fourth Meditation (Cottingham *et al.* 1984, 37 ff.). This basic picture of belief is apparently older than Descartes; Derk Pereboom traces it back to the Stoics in his (1994).

Of course, the Argument Argument might be filled out in a less contentious way. It might be interpreted not as a claim that arguments can have *no* cogency-independent effect on rational belief. It might claim instead that the graded-belief-mediated effects of arguments are *insufficient* to explain the role that arguments legitimately play in our epistemic lives. Is it plausible that arguments legitimately affect us epistemically *only* via their effects on our graded beliefs?

Given the enormous variety of arguments, and of accounts of binary belief, it is hard to say much in general about this question.[12] And it would clearly be impossible to prove that no case exists in which a deductive argument rationally affects binary belief in a way that cannot be explained via the argument's effects on rational graded belief. But we can, I think, see that in some very typical instances, ordinary deductive arguments will have dramatic effects on rational graded beliefs, and these effects are just the sort we would expect in turn to affect binary beliefs—and affect them in exactly the way we traditionally associate with reasoning deductively. For example, suppose I look at my office answering machine, and form a very strong belief that

(1) My office answering machine recorded a call as being from local number 865–4792 at 1:45.

I already have the following strong beliefs:

(2) My house is the only one with local phone number 865–4792.
(3) My wife, son, daughter, and I are the only ones who live at my house.
(4) If 1 and 2, then someone called from my house at 1:45.

[12] We may generalize our point about threshold-style accounts a bit, and note that any account that allows degrees of rational confidence to provide a floor for rational binary belief, or a ceiling for rational non-belief, will be sensitive to arguments' effects on graded belief. And any account that does not do this must allow belief in vastly improbable propositions, or allow non-belief in virtual certainties. But the lottery and preface cases have already shown us that this price must be paid by any account of belief which protects deductive cogency requirements.

(5) If 3, and someone called from my house at 1:45, then my wife, son, daughter, or I called from my house at 1:45.
(6) I didn't call from my house at 1:45.
(7) My son and daughter were at school at 1:45.
(8) If 7, then neither my son nor my daughter called from my house at 1:45.

I then form the following strong belief:

C: My wife called from my house at 1:45.

Here we have a valid deductive argument with eight premises, each of which is necessary for deriving the conclusion. Moreover, each of the premises is something we would, in our ordinary binary belief-attributing practice, describe me as believing. And my belief in C is based on, and made rational by, my beliefs in 1–8. This seems to be a paradigm example of the sort of deductive reasoning we engage in daily. Can we account for examples like this by means of the argument's effect on rational graded belief?

Let us first think about how strongly I would rationally believe the premises. With respect to premises 1, 2, 3, 4, 6, and 8, I am virtually certain of each. The chance of my being wrong on any of these counts is surely less than 1 in 1,000; so let us set my degree of confidence in each of these, very conservatively, at 0.999. With respect to 5, there is some possibility that a call from my house to my office would have been placed by someone who didn't live at my house. True, I've received hundreds of calls from my house, none of which have come from anyone who didn't live there. But suppose we allow a very generous 1% chance of 5 being false, and set my degree of confidence in 5 at only 0.99. With respect to 7, it is possible that one of my children has, e.g., become sick at school, and has been brought home by my wife. But this certainly happens way less than once a year (and when it does, my wife lets me know as soon as she is called). Again, let us be very conservative, and set my degree of credence in 7 at only 0.99.

Now, as we've seen, the fact that each of these premises is itself highly probable does not entail that I must give high probability to their conjunction. But in the present case, there's no reason to think that the truth of any of the premises provides much reason for me to disbelieve any of the others. Let us suppose that they also don't lend one another significant mutual support, so that they are mutually independent. (This is of course not strictly true, but I think that they are independent enough so that we will not err too greatly in treating them as if they are. It is important to remember that we were extremely conservative in our original credence-assignments.) On this assumption, my credence in the conjunction of premises should be somewhat greater than 0.974. And it seems to me that this rough calculation passes the intuitive test: in the present sort of case, I should be very highly confident that all of 1–8 are true. Given this, of course, it follows immediately that my credence in C must be at least this high.

Of course, one example—even if it seems fairly typical of our day-to-day reasoning with deductive arguments—cannot refute the claim that there are other cases in which graded-belief effects cannot explain the legitimate power of deductive arguments. In general, cases in which rationally persuasive arguments can be understood as operating through graded beliefs are likely to be cases of arguments where we are very confident of the premises, where we don't have too many premises, and where the premises are positively relevant to one another—when they form an integrated, mutually supporting structure of claims—or are at least not negatively relevant to one another. In cases of these sorts, it will be rational to have reasonably high confidence in the conjunction of the premises of the argument, and, therefore, in the argument's conclusion.

In other cases, it is undeniably true that the graded-belief-based effect will be negligible. This will happen, for example, in arguments with large numbers of fairly uncertain or mutually negatively relevant premises. In these cases, results in conformity with the dictates of deductive cogency cannot be shown to flow from the argument's effect on graded belief. But at least some of the clear cases in this category should give considerable pause to the advocate of the Argument Argument. For this category includes, paradigmatically, the very arguments where deductive cogency would lead us from reasonable premises to intuitively absurd conclusions: that my paper will never fail to be on my porch in the next year; that my history book is the very first error-free contribution to my field; that I, perhaps alone among all the people who have ever lived on Earth, believe only truths!

In response to the suggestion that the force of *reductios* depends on number of premises, Kaplan offers an example designed to show that even large *reductios* have rational force—force which, owing to the large number of premises, cannot be explained probabilistically.[13] Kaplan asks us to suppose that he's been asked to produce a chronology of 26 events leading up to a serious accident. The chronology he produces has the following elements:

P1: Event A preceded event B.
P2: Event B preceded event C.
P3: Event C preceded event D.
.
.
.
P26: Event Z preceded event A.

We point out that, given transitivity and non-reflexivity of temporal precedence (which he accepts), his chronology logically implies a contradiction. This *reductio*, Kaplan argues, "has critical bite: it exposes the fundamental inadequacy of the chronology I have produced."

It is clear that the imagined chronology is fundamentally inadequate; but there are many sorts of inadequacy. If the example is to demonstrate the efficacy of certain *reductios*, we must, in assessing it, be clear about which claim is the target of the *reductio*. Perhaps the

[13] Kaplan (2002: 459–60, fn. 20). Kaplan credits Ruth Weintraub and Scott Sturgeon for the suggestion to which he is responding.

most natural target to consider is the chronology itself, considered as one conjunctive claim. But this choice would demonstrate no problem at all; if the *reductio*'s bite is to render belief in this great conjunction irrational, we have no difficulty explaining this bite on the basis of low probability.

In fact, similar points apply to other intuitively salient targets. Reconstructing important aspects of the events leading up to an accident will undoubtedly make use of transitivities to arrive at judgments about the temporal relations between, e.g., event D and event J. But if the events in which we're interested are the ends of a seven-link chain (as D and J are), the probability of the temporal-priority judgment we would naturally reach about them (that D preceded J) is less than 3/4![14] And if the events we're interested in are the ends of a 13-link chain, the chronology gives us no reason to place greater credence in either one being prior. Thus, when we think about many useful claims that might naturally be taken to be part of the chronology, there again seem to be good probabilistic explanations for the chronology's intuitive inadequacy.

This suggests that, insofar as the case poses a difficulty, it is with the individual elemental claims such as

> P6: Event F happened before event G.

Indeed, these are the claims to whose probabilities Kaplan directs our attention: he points out that the situation described is compatible with our rationally having high (> 0.96) credence in each of these judgments. So perhaps the argument will be that although this level of confidence seems quite compatible with binary belief, the (large) *reductio* of P6 shows that P6 is not really belief-worthy. Since its unworthiness cannot be explained by the *reductio*'s probability-lowering effect, cogency is required to account for the rational effect of argument here.

[14] This assumes that we have no reason to trust some elements more than others, and that only one of the elements is false. If the events related by a particular judgment are the ends of a 7-link chain, then there's a 7/26 chance that the error is in one of those links (in which case the judgment is false); thus, there is only a 19/26 chance that it's true.

Now I think that it is not at all clear that P6 is unworthy of belief. But before thinking more carefully about this question, I'd like to fill the case out in a bit more detail, the better to fix our intuitions. First, the circular structure of the elements may be doing some intuitive work by suggesting that there is at least one *big* mistake in the elements. If that's right, we may well be influenced by the thought that the source from which we obtained evidence for the elements was not good, and that we shouldn't really be very confident of any of them. True, giving the elements 0.96 probability is mathematically possible in a case of this abstract structure. But that doesn't make 0.96 an intuitively realistic estimate of the probabilities in an actual case meeting the description. Let us, then, specify how the elements are arrived at. One might naturally imagine that the various events could somehow have been timed by relations to external events (e.g., the car was filled with gas at 8:15 am). But this would not lead to a circle of priority claims. To fix our intuitions as clearly as possible, let us try to fill in the abstract description in a fairly natural way, so that our evidence will lead to a circular structure with the high probabilities the argument requires. Here's one way of doing so (I've also taken the liberty of changing subject matter, to remove any distraction that might be caused if our supposition that we're reconstructing a serious accident for some important purpose had the effect of raising the intuitive standard for rational belief above 0.96).

Suppose there is a 26-person race, which we haven't seen. The rules stipulate that each racer will tell us who finished right behind her (and will tell us nothing else). The rules (which we may suppose are followed religiously) also stipulate that all racers tell us the truth, with the exception of the last finisher, who is to tell us that the racer who actually won finished behind her. We thus arrive at intuitively reasonable probabilities of 0.96+ for the elemental claims such as "racer C finished before racer D." Here, it is even less clear that belief in these claims would be irrational. Nevertheless, I think it must be acknowledged that many would be hesitant to assert unqualifiedly that D finished before E. And I think that some would also deny the rationality of believing that D finished before E. (My own intuition, though not strong, is that this belief would be rational; but let us put that aside.) Supposing that we hold that such beliefs would not be rational, this could not be explained on probabilistic grounds. Would this show that there was, after all, a need to invoke cogency to explain why the beliefs weren't rational?

It seems to me that no such conclusion would be warranted. For once we have filled out the epistemic situation to rationalize the high probability judgments for the elemental beliefs, and once we have focused our attention on these elemental beliefs rather than on certain other beliefs that might be derived from them, the case very much resembles a standard lottery case. And this suggests that cogency demands may not be the best explanation of our reluctance to attribute rational beliefs. To test this suggestion, let us consider another race case, this one modified to remove the threatened *reductio*.

Suppose we receive 26 reports on who won each of 26 independent two-person races (e.g. "D beat K in race 6"). Suppose, however, that some of the people who report results are less than perfectly reliable—in fact, we know that, over a very long run, score-reporters have lied (with no discernible pattern) 1/26th of the time in this sort of context. Knowing this, what should we say if someone asks us who won race 6? I think that many would be reluctant simply to assert unqualifiedly that D beat K. And I think that those who were reluctant to countenance rational belief that racer D preceded racer E in the previous case would likely be just as reluctant to countenance rational belief that D beat K here.[15]

[15] It is also worth remembering that, if the defender of cogency claims that one should believe the elements in the second race case, he must also hold that one should believe that, for some reason, the score-reporters told us the truth 26 times in a row! This is highly counterintuitive already (since on anyone's account, they probably haven't made 26 reports without lying); moreover, it is hard to see any reason for saying this in the present case that would not apply even if the reporters had given us 260 reports. Thus, insisting on rational belief in the second race case does not seem like an attractive option for the defender of cogency.

If that is correct, then it seems that the failure of rational belief in these two race cases should be explained in the same way. And the explanation does not flow in any obvious way from cogency, since in the second race case there is no *reductio*—no guarantee that one of the elements will be false. (This is related to the point that our reluctance in lottery cases to assert flatly, or avow belief in, claims such as "ticket 17 won't win" is not diminished when the lottery isn't guaranteed to have a winner.) Thus it seems to me that in the end we still have not seen a case in which the legitimate effect of argument on rational belief needs to be explained by a demand for deductive cogency.

Does this show that there are no examples that would serve the purpose of the Argument Argument? Certainly not. But those who would question deductive cogency requirements surely cannot be expected to demonstrate exhaustively that in every case where a deductive argument affects binary belief in an intuitively legitimate way, this effect can be explained independently of cogency. Surely the burden is on proponents of the Argument Argument to come up with specific, detailed examples of arguments whose rational efficacy cannot be explained in cogency-independent terms. For as we have seen, the general point that deductive arguments play a crucial epistemic role for us does not in itself establish a role for deductive cogency requirements.

It might be objected that I've underplayed the seriousness with which we actually take inconsistencies in the context of inquiry. Suppose, for example, the author of a history book were to discover that the claims *in the body of her book* formed an inconsistent set. Intuitively, wouldn't this be very disturbing? Might the fact that the crucial Modest Preface claim is, in some sense, "not really about history anyway"—that it oversteps, in some intuitive sense, the context of inquiry—make preface-type inconsistencies seem acceptable?

Now as we've seen, one can make a preface-like point with a great conjunction of the purely historical claims in the body of the book. And even bracketing this point, it is hard to see why an author *should* be more concerned by an inconsistency within the body of the book than with preface-style inconsistency. After all, our comfort with the Modest Preface statement is directly based on our being highly confident that at least one of the claims in the body of the book is false. Discovering that the claims in the body of the book form an inconsistent set may elevate that high degree of confidence to certainty, but it is hard to see why this slight increase in our degree of confidence should be so alarming.

But wouldn't discovering inconsistency among the individual historical claims in the body of the book always *actually be* highly disturbing? I think that the answer to this question is less clear than it might seem at first. What the defender of cogency needs to make his point is a case involving an inconsistency that necessarily involves a great number of the huge and diverse set of historical claims making up the body of a book, and for my part I know of no case in which we've had experience of this sort of discovery in actual inquiry. Undoubtedly, people have found inconsistencies in the bodies of books, where the inconsistencies have been generated by a fairly small number of claims. But as we have seen, graded-belief-based effects may explain our felt need for revision in this sort of case. And in certain other cases, discovery of an inconsistency impugns one's general methods or sources in a way that significantly reduces one's confidence in some or all of the particular claims in the book. Again, however, our being disturbed in such cases can be explained in degree-of-belief terms. The kind of example that would bolster the argument for cogency would have to be one in which the discovery of the inconsistency did not significantly lower our confidence in the truth of any of the book's claims. Lacking experience with such cases, we cannot assume that they would actually strike us as calling urgently for epistemic repair.

Until persuasive specific examples are found, then, it seems to me that we've been given no good reason to think that deductive cogency requirements play an important part in epistemic rationality. Moreover, I think that at present we have at least some reason for skepticism about the prospects for finding examples that will suit the Argument Argument's purposes. For any such example will have to be one in which we think that it is rational for someone to believe the conclusion of an argument based on the argument's premises, where all the premises are necessary to reach the conclusion, and yet where we also think that it's *not* rational for her to be confident that the premises are all true![16]

Finally, it should be kept in mind that success for the Argument Argument would not be secured even by the discovery of a few cases that seem intuitively to fit this description. For the argument's success, there would have to be a considerable range of such cases. After all, aside from any intrinsic implausibility of the claim that rational beliefs may be based on premises that the agent is rationally quite confident are not all true, our intuitive verdicts on many cases provide powerful reasons to reject the demands of cogency. Thus, the cases adduced in support of the Argument Argument would have to be pervasive and persuasive enough to counterbalance the intuitive absurdities entailed by cogency requirements in the preface case, in newspaper-type cases, and in our ordinary expressions of epistemic humility.

In sum, then: there is certainly considerable surface plausibility to the idea that deductive arguments must derive their epistemic bite from deductive cogency requirements on binary belief. But it is also plain that submitting binary belief to cogency leaves us subject to bizarre arguments which run roughshod over our common-sense

[16] This is not to beg the question by arguing that an intuitively persuasive example fitting this description should be disregarded just because it fits the description. It is intended merely to point out that examples of arguments whose effects cannot be accommodated by graded-belief-based mechanisms are going to resemble the strikingly counterintuitive applications of cogency in some respects—respects that are likely to make them counterintuitive as well. One might object that our intuitions in such cases would be distorted by our confusing binary belief with a state of confidence. But the objection would itself beg the question if it meant to argue that any intuitions based on rational graded belief must be discarded; after all, the degree to which rational binary belief depends on rational graded belief is part of what is at issue. If there are no cases in which an argument affects binary belief in a way that is very clearly correct intuitively, yet which cannot be explained via graded-belief effects, the Argument Argument is a non-starter.

understanding of rational belief. Insofar as there is an alternative way of grounding those deductive arguments which are intuitively legitimate contributors to epistemic rationality, we may resolve this tension nicely: we may maintain a healthy respect for rational argument without capitulating to the exorbitant demands of deductive cogency.

4.4 Rational Binary Belief

We customarily talk, and think, about our beliefs in binary terms. And it is certainly plausible to say that the point of beliefs is to represent the world accurately; that one's beliefs should comprise as much of the whole true story of the world as possible; that deductive arguments play an important role in determining which beliefs it is rational to have. But none of these observations about binary belief turn out to provide a sound motivation for a cogency requirement on binary belief.

Now this does not show that there is nothing to binary beliefs, or that there is no purpose to our talking about beliefs in an all-or-nothing way. It is clear that our everyday binary way of talking about beliefs has immense practical advantages over a system which insisted on some more fine-grained reporting of degrees of confidence. This is clear even if binary beliefs are understood on a simple threshold model.[17] At a minimum, talking about people as believing, disbelieving, or withholding belief has at least as much point as do many of the imprecise ways we have of talking about things that can be described more precisely.

To take a trivial example, consider our practice of talking about dogs as big, small, and medium-sized. Obviously, talking about dogs in this way is extremely useful in everyday contexts. We would not

[17] Foley, who defends a threshold model, makes a convincing case for the utility of binary belief-talk (Foley 1993, 170 ff.). Weintraub (2001) defends a threshold view along similar lines.

want to deny that, in a perfectly straightforward way, some dogs are big and some aren't, even if more precise ways of talking about dog sizes are available. And our rough sorting of dogs into three sizes even figures in everyday explanations: Andy provides good protection because he is big; Sassy is cheap to feed because she is small; etc. No one would advocate wholesale replacement of our everyday way of talking and thinking about dog sizes by some more precise metric—say, in terms of weights (or heights, or approximate weights or heights, or some function of approximate weights and heights). Any such wholesale change would clearly be counterproductive.

Nevertheless, as the example suggests, the obvious usefulness of talking about things using a given category doesn't show that the category "cuts nature at its joints." In the dog-size case, the interesting regularities—even the ones underlying the explanations mentioned above—will be more likely to be framed using more precise metrics. Small dogs do tend to eat less than big ones; but this regularity itself is explained by the way in which food consumption tends to increase with size, even within the "small dog" range. When we get serious about size-dependent effects—e.g. in calculating dosages of medicine—more precise metrics are quickly employed. Our rough-and-ready size categorizations do not seem to reflect the fundamental structure of the phenomena they describe.

Does our ordinary binary way of talking about beliefs pick out some epistemic property that's more important than bigness in dogs? Many epistemologists—even those, such as Foley, Maher, and Kaplan, who see graded beliefs as playing an important epistemic role—seem to think so. Kaplan, for example, considers a case in which you've just reported exactly how confident you are that a certain suspect committed a crime:

> One of your colleagues turns to you and says, "I know you've already told us how confident you are that the lawyer did it. But tell us, do you *believe* she did it?" (Kaplan 1996, 89)

For Kaplan, there is something epistemically important left out when we give a description of a person's degrees of confidence.

For my own part, the colleague's question feels a lot like the question "I know you've told us that the dog weighs 79 pounds and is 21 inches high at the shoulder. But tell us: is it big?" When I "enter most intimately into what I call *myself*," I find no discrete inner accepting or "saying yes" to propositions. This seems particularly clear in cases where I move gradually from a state of low credence to a state of high credence (or vice-versa). I may start a picnic having heard a very positive weather forecast, and having no reservation about saying "I believe we're going to have a great picnic." But during the course of an hour, as clouds appear on the horizon and move toward us, as the sky gradually darkens, and as the breeze becomes stronger, my confidence in having a pleasant time fades, through the point where I no longer would self-ascribe the belief that we're going to have a great picnic, until, at the end of the hour, I would unhesitatingly say "I believe that our picnic is going to be spoiled." But at no point during the process do I seem to experience a discrete qualitative shift in my attitude toward the proposition that we'll have a great picnic—no jumps from an inner "saying yes" to an inner withholding of judgment to an inner "saying no." If, at some point in this process, I had said that I thought that the chances of our picnic being spoiled were 9 to 1, and someone asked, "But do you *believe* that our picnic will be spoiled?" I quite literally would not understand what information she was asking for.

Nevertheless, I don't mean to suggest that our binary belief talk is governed merely by degrees of confidence. As we've seen, we are somewhat reluctant to attribute beliefs in cases where the agent's high degree of confidence is based on blatantly statistical grounds. Various explanations of this fact may be offered: perhaps our belief-attribution practices are sensitive to some explanatory or tracking requirement, or to the fact that the statistical grounds somehow render salient the possibility of having the same evidence while being wrong. Adjudicating among these explanations would be a substantial project in itself.[18] But it is far from obvious that finding the right explanation—some rule or aspect of our belief-attributing practice that disqualified blatantly statistically based beliefs—would help reveal some binary state that was subject to interesting rational constraints (beyond those affecting degrees of confidence). This seems especially evident if our belief-attributing practice turns out to be sensitive to contextually determined conversational saliencies.

To take another example, as Nozick (1993, 96–7) suggests, our willingness to attribute belief may depend on what practical matters are at stake. I would unhesitatingly describe myself as believing that our picnic will be a success if I were 98 percent confident that it would be a success and 2 percent confident that it would be spoiled by rain. But if I were only 98 percent confident that our airplane would arrive safely and 2 percent confident that it would crash, I would not unhesitatingly describe myself as believing we'd arrive safely. Again, various explanations of our practice are possible. For example, attribution of a belief that P might require the agent to have a high degree of confidence in P, but what counts as high might be sensitive to how badly wrong things could go if P is false. But again, without adjudicating among possible explanations, we can see how a practice of making black-and-white belief-reports that are sensitive to factors beyond degrees of confidence might make perfect sense, without its revealing any rationally interesting underlying epistemic state going beyond degree of confidence.

The project of working out the conditions under which people appropriately attribute binary beliefs may well reveal an interesting and complex pattern, even if our belief-attributing practice does not in the end correspond cleanly to a kind of state that is important from the point of view of epistemic rationality. After all, even the conditions under which we call dogs "big" may be interestingly

[18] Writing about our willingness to make unqualified assertions, Kaplan comments: "Why we discriminate in these ways against matters of chance I am at a loss to say, but that we do seems quite clear" (1996, 127).

complex. Our practice there may not be governed by any fixed threshold of weight or height or weight/height combination. Perhaps factors such as the average size in some contextually relevant subset of all dogs, or the contextually specified use to which a particular dog is to be put, help determine our judgments. But working out complexities of this sort would not, I think, disclose any property that was important from the point of view of a systematic study of canine sizes.

The general reason for worrying that binary belief will not turn out to be an important part of epistemic rationality is this: insofar as our binary belief-attributing practices are sensitive to factors beyond rational degree of confidence in a proposition's truth, those practices are likely to point away from what we are most concerned with when we think about epistemic rationality. Let me illustrate with one clear example of this tension between going beyond rational degrees of confidence and maintaining epistemic importance. We saw earlier that BonJour (1985), in response to standard lottery cases, denies that one is fully justified in believing that one's ticket will lose, no matter how high the probability is (though one may be fully justified in other beliefs whose probabilities are lower). But this move—exactly the sort needed by defenders of deductive cogency—seems to run directly counter, at least in spirit, to BonJour's own characterization of epistemic justification:

> [A]ny degree of epistemic justification, however small, must increase to a commensurate degree the chances that the belief in question is true (assuming that these are not already maximal), for otherwise it cannot qualify as epistemic justification at all. (BonJour 1985, 8)

The worry is that there is no interesting notion of epistemic rationality that will sanction an agent's believing P but not sanction her believing Q, in a situation when she rationally believes that Q is more likely to be true than P.

This theoretical worry, of course, applies to virtually any bifurcated concept of binary belief. But if one also insists that rational binary beliefs be *deductively cogent*, then worries about the significance of belief so understood multiply. Consider, for example, Professor X, our deductively cogent historian. We've seen how his belief in the Immodest Preface Proposition will commit him, given certain quite unexceptionable background beliefs, to believing some quite remarkable things: that he'll soon be receiving opportunities for professional advancement, that in one year he'll be enjoying a handsome salary and driving a brand-new Alfa-Romeo, etc. In Section 3.4, we saw that the intuitive irrationality of these beliefs (there labeled (a) through (e)) makes them prima facie counterexamples to cogency requirements. Here, I'd like to highlight a somewhat different angle: supposing that these *are* examples of rational binary beliefs, what do they reveal about the species of belief they exemplify?

Let us first think about how Professor X's beliefs should relate to the practical decisions he'll be making. Suppose, for instance, that he is offered an excellent deal on a new sensible car. His present sensible car could be nursed along for another year, so buying now will be quite disadvantageous if he buys a new Alfa-Romeo one year hence. On the other hand, if he does not receive the infusion of cash that would make the Alfa possible, he will do much better by taking advantage of the present offer. He believes, of course, that he'll be buying an Alfa in one year. Should he turn down the good deal on the sensible car? Obviously, he should not. The binary belief that he'll be buying the Alfa in one year, like various other beliefs that flow from the Immodest Preface belief, must be walled off carefully from Professor X's practical reasoning, lest he be led into countless idiotic practical decisions.

I should emphasize that defenders of deductive cogency requirements have sometimes said quite forthrightly that only graded belief should figure in practical deliberation. But this position seems much more palatable when one concentrates on just the Immodest Preface Proposition, whose obvious and direct practical implications are minimal. When the belief that one will be buying an Alfa in one year gets disconnected completely from the practical question of whether to buy a car now, the point in having such a belief comes into question.

Moreover, disconnecting beliefs from practical reasoning in this way has bizarre implications for what one should *believe* about what one has practical reason to do. Suppose Professor X believes, as it seems he should, that

(f) Anyone who has a perfectly decent car and is going to buy a new car in one year should not buy a new car now.

Given his beliefs about his own situation, deductive cogency would have him believing, quite rationally, that

(g) I should not buy the new car now.

But this verdict must somehow cohere with the obvious fact about practical reasoning noted above: that Professor X would be quite irrational not to buy the new car now!

Moreover, the problem is not just that the deductively cogent agent's beliefs about his reasons for action are prised so far apart from what he actually has reason to do. A bit more exploration of these beliefs themselves raises serious doubts about the very coherence of cogency's demands. It seems obvious, for example, that Professor X should believe

(h) It's very *unlikely* that I'll be able to afford an Alfa in the next few years.

But it's also hard to deny that he should believe

(i) If it's very unlikely that I'll be able to afford an Alfa in the next few years, I should buy the new car now.

And given these beliefs, cogency would require Professor X to believe

(j) I should buy the new car now.

The problem here is not, of course, that (j) is intuitively irrational—quite the opposite is true. The problem is that belief in

(j) is also *prohibited* by cogency, given Professor X's (cogency-mandated) belief in (g). It is not obvious just what beliefs a defender of cogency should recommend in this situation. Unless some way is found to deny the rationality of the beliefs leading to (g) or (j), it seems that cogency turns out to be unimplementable.

One way of avoiding this difficulty might be to argue that assertions and self-attributions of belief made in the context of practical decision-making did not express binary belief after all, but instead expressed degrees of confidence. I won't attempt to work out the intricacies of such an approach here. But I will note that, if there were a suitable way of circumscribing contexts of practical decision-making, it would effect a further corralling of the sort of belief to which cogency applied: such beliefs would end up being separated even from our ordinary ways of *thinking* and *talking* about practical decisions. (It's worth noting that this corralling of binary belief would have to exclude it even from certain contexts of inquiry. It's obvious that scientists, historians, etc., must make practical decisions in conducting their work—for example about expending research effort. But beliefs relevant to such decisions are subject to the sort of problem embodied above in Professor X's beliefs about whether he should buy the new car.)

In addition, there are many other ways in which the beliefs mandated by deductive cogency must be isolated from central parts of the agent's life. Let us ask: should Professor X be happy and excited that he will soon be enjoying a handsome salary and giving prestigious talks? Should he be surprised when he doesn't win the SHE prize? Presumably not. So, while it is obvious that one's emotions should in general be responsive to one's beliefs about the world, it is equally clear that they should not be responsive—at least not in any intuitively attractive way—to the sort of binary beliefs that would satisfy deductive cogency.

Now it might be claimed that disconnecting binary beliefs from emotive aspects of an agent's cognitive system is not much of an additional price to pay, once one has disconnected the beliefs from the agent's practical reasoning; after all, one might expect both practical reason and emotions to be closely interconnected through the agent's values. But one might well think that if binary beliefs are to have any importance at all they cannot end up being a wheel that turns nothing in the cognitive life of the agent. What is left? As we have seen, one connection that is explicitly advocated by Kaplan is with the agent's assertive practice. Can deductively cogent binary belief find at least some of its importance in its connections with assertion (or, more specifically, with assertions motivated by the aim of truth-telling)?

It seems to me that even this connection is quite dubious. As we saw above, the rational assertability of the Immodest Preface statement was already intuitively suspect. It would be ridiculous (and, given how we ordinarily interpret assertions, even dishonest) for Professor X to assert in an NEH grant application that he had written the first error-free book in his field. And thinking about beliefs "downstream" from the Immodest Preface belief makes even clearer the strain that would accompany systematically connecting cogency-regulated belief to assertions. Consider assertions about the future. Should Professor X (insofar as he wants to tell the truth about the matter) assert to his friends that he'll soon be driving an Alfa-Romeo? Intuitively, it seems not—such an assertion would be highly misleading.

The connection is strained further when we consider what Professor X should assert about the practical decisions he or others need to make. Would it be reasonable for him to assert sincerely to a graduate student that trying to find even small errors in his book would be a waste of the student's time? Should he assert that he himself should not buy a new car now (even as he quite reasonably buys one)? Again, it does not seem that the kind of binary beliefs mandated by deductive cogency provide a basis for reasonable sincere assertion.

It turns out, then, that thinking about the sorts of binary beliefs required by cogency in some quite ordinary circumstances reveals strong reasons for doubting the importance of cogency-respecting binary belief. The reasons go beyond the general theoretical worry about any bifurcated notion of belief: that such belief is determined by factors insufficiently related to truth. They concern the difficulty of connecting this sort of belief in any intuitive way with the rest of the agent's concerns, attitudes, or practices. To put the point another way, examples like those considered above raise the following sort of question: what point would there be in a practice of selecting a favored set of propositions to "believe," if this set of propositions included propositions of the sort Professor X is required to believe by deductive cogency?

Again, this is not to deny that our practice of binary belief-attribution is useful: clearly, it is. Moreover, it might well be interesting to see what governs this practice, and in so doing to gain insight into what we're talking about when we attribute binary belief. What is somewhat doubtful, though, is that this project will reveal to us a species of belief that will prove important from the point of view of epistemic rationality. And if we take binary belief to be a state that is governed by the constraints of deductive cogency, doubts about the state's epistemic importance become particularly acute. So while the ultimate interest of binary belief remains open to debate, it seems to me that if logic has a role to play in shaping epistemic rationality, it will not be the traditional one of subjecting binary belief to deductive cogency.

5 LOGIC, GRADED BELIEF, AND PREFERENCES

5.1 Graded Beliefs and Preferences

THE suggestion that logic contributes to epistemic rationality primarily through imposing conditions on graded beliefs is a relatively new one in the history of thinking about logic. But we've already seen that the traditional approach of imposing deductive cogency on binary belief, despite its undoubted intuitive naturalness, cannot capture the way logic informs epistemic rationality. Moreover, we've seen that what is perhaps the central role that logic has traditionally been thought to play in our epistemic lives—subjecting rational belief to valid argument—may be explained not by a cogency requirement on binary belief, but instead by constraints on rational degrees of belief. For these reasons, it is worth taking seriously the possibility that logic gains its epistemic purchase on us primarily through the constraints of probabilistic coherence.

The idea that probabilistic coherence is a rational requirement—let alone the primary way that logic informs epistemic rationality—has, however, met with quite a bit of resistance. Some of the resistance stems from the impression that the mathematics of probabilistic coherence involves an unacceptable level of idealization: it just seems wrong to suppose that we accord mathematically precise probabilities to the various propositions we have beliefs about—or even, many would hold, that it would be ideal to do so. I'd like to put off discussion of the role of idealization in epistemology for now, though, to concentrate on a more fundamental source of resistance to probabilistic coherence requirements. This source of resistance stems from the fact that proponents of probabilistic coherence have traditionally cast their arguments in a way that makes their subject matter—graded belief—seem much less like binary belief than one might at first have supposed.

Let us call the view that ideally rational degrees of belief must be probabilistically coherent "probabilism." The traditional arguments for probabilism have tried to accomplish two tasks simultaneously. The first—a quasi-descriptive or stipulative task—is to provide for some way of defining and/or measuring graded beliefs. This has seemed necessary in part because our natural way of thinking and talking about beliefs is binary; graded beliefs seem in a way more like "theoretical" entities than like common-sense objects of our everyday epistemic experience. The second task the traditional arguments have sought to accomplish is a normative one: to show that graded beliefs, so defined, should be probabilistically coherent. Both of these tasks have been accomplished by tying graded beliefs to something that is not obviously within the epistemic realm: preferences. Degrees of belief are *defined* in terms of preferences, and then intuitively rational conditions on preferences are shown to impose probabilistic coherence on these degrees of belief.

The obvious worry occasioned by such arguments is that we've strayed from the topic of epistemic (as opposed to pragmatic) rationality. And this worry is sharpened by the fact that our natural way of isolating epistemic rationality invokes a goal of something like accurate representation of the world. This has an obvious application to binary beliefs; after all, the propositions we accept can be true or false, and accurate representation of the world can naturally be thought of in terms of believing true propositions and not believing false ones. But there is no similarly obvious sense in which, say, believing a true proposition to degree 2/3 contributes to the accuracy of the agent's representation of the world.[1]

[1] This is not to say that there is no way of capturing this idea; in fact, various proposals have been advanced for characterizing the accuracy or nearness-to-the-truth of graded beliefs. James M. Joyce (1998) has even shown, for certain attractive measures

Some advocates of pragmatic approaches to graded belief have been sanguine about the thought that defining graded beliefs in terms of preferences makes them into something quite unlike the beliefs we wonder about pre-theoretically. Richard Jeffrey, for example, endorses Ramsey's idea that the state we define in terms of an agent's preferences is the agent's "belief *qua* basis of action." Jeffrey writes:

> [I am not] disturbed by the fact that our ordinary notion of *belief* is only vestigially present in the notion of degree of belief. I am inclined to think that Ramsey sucked the marrow out of the ordinary notion, and used it to nourish a more adequate view. (Jeffrey 1970, 171–2)

It seems to me, however, that this sanguinity is misplaced. For one thing, the move of defining degrees of belief in terms of an agent's preferences (as revealed in her choice-behavior) is reminiscent of the standard operationalist strategy in philosophy of science: taking one way of measuring a theoretical quantity and treating is as a definition. Bruno de Finetti, one of the founders of the preference-based approach to graded belief, is quite straightforward about his operationalist motivations in this matter. Commenting on his definition of personal probabilities in terms of betting preferences, he writes:

> The important thing to stress is that this is in keeping with the basic requirement of a valid definition of a magnitude having meaning (from the methodological, pragmatic, and rigorous standpoints) instead of having remained at the level of verbal diarrhoea... (de Finetti 1977, 212)

But today, operationalism and kindred approaches to theoretical magnitudes are widely seen to be misguided. And this goes not only

of accuracy, that any set of graded beliefs that violates the probability axioms can be replaced by a probabilistically coherent set that is guaranteed to be more accurate. Joyce offers this as a clearly non-pragmatic vindication of probabilism. Unfortunately, as Maher (2002) has pointed out, there are other accuracy measures that do not support this result, and the arguments in Joyce (1998) that would rule out these measures are not fully convincing. At this point, it seems to me that the jury is still out on the prospects for providing a clear accuracy-based argument for probabilism. See Fallis (2003) for useful further discussion and references related to this topic.

for physical quantities such as length and temperature, but also for psychological concepts such as pain, intelligence, and belief.

Unfortunately, the traditional arguments supporting probabilistic coherence as a norm for graded belief make explicit use of definitional connections between beliefs and preferences. This raises the question: can we support probabilistic coherence as a norm for rational degrees of confidence, without making graded beliefs into something that they are not? Clearly, the answer will depend on the way in which the preference-based definitions figure in the relevant arguments. In this chapter, I'll look more closely at the two main strands of preference-based argument that have been used to support probabilistic coherence requirements: Dutch Book Arguments, and arguments based on Representation Theorems.

5.2 Dutch Book Arguments and Pragmatic Consistency

"Dutch Book" Arguments (DBAs) are the best-known way of supporting the claim that one's graded beliefs should be probabilistically consistent. The arguments' central premise posits a close connection between an agent's graded beliefs and her betting behavior: the agent's degree of belief in a proposition P is assumed to be measurable by her preferences as they are expressed in her willingness to accept bets on P. Though the details of the betting arrangements in various DBAs differ somewhat, they all involve the agent accepting bets at the odds dictated in the intuitively natural way by her degrees of belief. For example, on the basis of my 0.75 degree of belief in my having sausages for dinner tonight, I would be willing to accept a bet at 3:1 odds that I will eat sausages, and equally willing to accept a bet at 1:3 odds that I will not have sausages.[2]

[2] In general, an agent's degree of belief in a proposition P is taken to be given by her *betting quotient* q. An agent's betting quotient for P is q if she would be indifferent between taking either side of a bet on P at odds of $q:(1-q)$. This general pattern fits

Of course, the agent's degrees of belief so measured may not obey the laws of probability—there may be no probability function that matches the agent's degree of belief function for every proposition about which the agent has a degree of belief. That will be the case if, for example, my degree of belief in P is greater than my degree of belief in $(P \vee Q)$. The DBAs show that in all such cases the agent will be willing to accept a set of bets on which she is guaranteed to lose money overall—no matter what the truth is about the matters on which the bets are made.[3]

The vulnerability to this sort of guaranteed loss is taken to indicate irrationality, and thus the lesson of the DBAs is supposed to be that ideally rational degrees of belief must conform to the probability calculus.

Now the key argumentative move—from the hypothetical vulnerability to guaranteed betting losses to constraints on rational belief—has seemed to many a *non-sequitur*. It has been pointed out, for example, that there are no clever bookies who know my degrees of belief and can compel me to wager with them. Clearly, Dutch Book vulnerability is not a real practical liability. Moreover, even if probabilistically incoherent agents were subject to real practical difficulties, it would not obviously follow that their beliefs were defective from the *epistemic* standpoint (as opposed to being merely pragmatically unwise).[4]

Defenders of the arguments have replied that the point of DBAs is not to indicate a practical problem. Rather, Dutch Book vulnerability indicates a kind of *inconsistency*. It is the inconsistency, not

the example in the text; 3 : 1 odds are the same as 0.75 : 0.25 odds. Thus, the agent is taken to have a degree of belief function that assigns a number from 0 to 1—corresponding to the agent's betting quotient—to each proposition about which she has beliefs.

[3] I will not rehearse the mathematical details of the proof that violations of the probability calculus entail Dutch Book vulnerability. The classic presentations are in Ramsey (1926) and de Finetti (1937). Prominent contemporary presentations include Skyrms (1975), Horwich (1982), and Howson and Urbach (1989).

[4] I have mentioned some representative criticisms, but there are more. For useful discussion and references to the literature, see Eells (1982), Maher (1993), Kaplan (1996), and Armendt (1993).

the likely prospect of monetary loss, that is the problem. This is an especially appealing kind of answer if one would like to see the probabilistic laws as, in Ramsey's words, "an extension to partial beliefs of formal logic, the logic of consistency" (1926, 41).

This general line of thought has considerable appeal; for although the DBAs have seemed persuasive to many, it is hard to see how they would have any force at all if their point were to reveal some practical disadvantage that came from violating the rules of probability. The suggested approach avoids seeing DBAs as crudely prudential. Rather than taking probabilistic coherence as an economically useful defense against being impoverished by transactions with improbably clever bookies, it sees probabilistic incoherence as involving structural defects in the agent's cognitive system.

On close inspection, however, the "inconsistency" that Dutch Book defenders are talking about is less parallel to standard deductive inconsistency than one might have hoped. The classic formulators of DBAs, Ramsey and de Finetti, did not simply make the assumption that certain degrees of belief could naturally be expected to lead to certain betting preferences: rather, they *defined* degrees of belief in terms of betting preferences. If degrees of belief are, at bottom, defined in terms of preferences, the inconsistency involved in having probabilistically incoherent degrees of belief turns out to be an *inconsistency of preference*. Thus, Ramsey writes:

> Any definite set of degrees of belief which broke [the laws of probability] would be inconsistent in the sense that it violated the laws of preferences between options, such as that preferability is a transitive asymmetrical relation . . . (Ramsey 1926, 41)

More recently, Brian Skyrms put the point this way:

> Ramsey and De Finetti have provided a way in which the fundamental laws of probability can be viewed as pragmatic consistency conditions: conditions for the consistent evaluation of betting arrangements no matter how described. (Skyrms 1980, 120)

Clearly, this sort of consistency of *preference* is not the sort of consistency one would initially expect to come from generalizing the notion of deductive consistency to degrees of belief.[5] Let us call this the "pragmatic consistency" interpretation of the DBAs.

It seems to me that there is something very unsatisfying about the DBAs understood in this way. How plausible is it, after all, that the intellectual defect exemplified by an agent's being more confident in P than in (P ∨ Q) is, at bottom, a defect in that agent's *preferences*? It is only plausible to the extent that we take seriously and literally the proposal that particular degrees of belief are defined by particular preferences—or, perhaps more precisely, that degrees of belief reduce to (or necessarily include) certain preferences. Now this proposal may not represent the considered judgment of all defenders of the pragmatic consistency interpretation of DBAs, some of whom also talk of the relation between beliefs and preferences in more ordinary causal terms. But the important point is this: for inconsistency in beliefs to *be* inconsistency of preference, certain preferences must *be* (at least a necessary part of) the beliefs.[6]

This seems at best a very dubious metaphysical view. It is true that one need not be an old-fashioned operationalist to hold that there is some constitutive connection between beliefs and preferences. Certain more sophisticated contemporary approaches to philosophy of mind—various versions of functionalism—still posit a deep metaphysical connection between beliefs and their

[5] Indeed, one might well doubt that "inconsistent" is the best word to use in describing preferences that violate transitivity, for example. Since this terminology has become established, though, I will for convenience continue to use the term in a broad and informal way.

[6] Some presentations of Dutch Book results simply assume that agents' betting preferences correspond to their degrees of belief (see Skyrms 1990). For explicit identifications/reductions/definitions of graded beliefs in terms of betting preferences, see de Finetti (1977); Ramsey (1926, 36); and Jeffrey (1965*b*, 1991). Howson and Franklin (1994) and Howson and Urbach (1989) identify an agent's degrees of belief with the betting quotients she takes to be fair (though they don't take these as entailing any willingness to bet). For interesting expressions of looser connections between beliefs and preferences, see Ramsey (1926, 30–35) and Armendt (1993, 7).

typical causes and effects (including other mental states such as preferences and, of course, other beliefs). But the causal interconnections that are said to define or constitute a belief are quite complex. They never simply require that a certain belief state necessarily give rise to certain preferences. This brings up a revealing tension in the pragmatic consistency approach to DBAs.

Suppose that beliefs are individuated—with respect to degree as well as content—by their causal roles. Then it might be that my high degree of belief that P is in a sense partially constituted by my belief's connections to, e.g., the fact that I would pay a lot of money for a ticket that is good for a big prize conditional on P's truth. But if beliefs are individuated by their causal roles, they will be individuated not only by their connections to particular betting preferences, but also by their connections to other psychological states—in particular, to other beliefs. If that is true, however, then my strong belief in P would also be partially constituted by its connections to my strong belief that (P ∨ Q).

This is where the tension comes in. The entire interest of taking the probability calculus as a normative constraint on belief depends on countenancing the real possibility that the second sort of connection might fail to measure up to probabilistic correctness: I might strongly believe P but not have a sufficiently strong belief in (P ∨ Q). But once we countenance this possibility, do we have any justification for refusing to countenance the following possibility: that I strongly believe P but do not have a sufficiently strong preference for receiving a prize conditional on P's truth? It seems to me that we do not. We have been given no reason to think that having certain appropriate betting preferences is somehow more essential to having a given belief than having appropriate other beliefs is. Thus, the interest of taking the probability calculus as a normative constraint on beliefs is predicated on countenancing the very sort of possibility—failure of a given belief to give rise to the appropriate other psychological states—that undermines the reductionism at the heart of the pragmatic consistency interpretation. An acceptable interpretation of the DBAs must acknowledge that partial beliefs may, and undoubtedly do, sometimes fail to give rise to the preferences with which they are ideally associated.[7]

It is important to note that these considerations do not undermine the view that theorizing about degrees of belief requires that we have some fairly reliable method—or better, methods—for measuring them. Nor do they undermine the view that eliciting preferences in certain ways can provide very reliable measurements of beliefs. But they do, I think, serve to break the definitional link on which the pragmatic consistency version of DBAs depends: they undermine the oversimplified metaphysical *reduction* of beliefs to particular betting preferences.

Rejecting this sort of reduction has an important consequence for the interpretation of DBAs. The arguments' force depends on seeing Dutch Book vulnerability not as a practical liability, but rather as an indication of an underlying inconsistency. Once we have clearly distinguished degrees of belief from the preferences to which they ideally give rise, we see that inconsistency in degrees of belief cannot simply be inconsistency of preferences. If the DBAs are to support taking the laws of probability as normative constraints on degrees of belief, then Dutch Book vulnerability must indicate something deeper than—or at least not identical to—the agent's valuing betting arrangements inconsistently.

Now one possibility here is to defend what might be called a "mitigated pragmatic consistency interpretation." One might

[7] A similar problem applies to a somewhat different consistency-based interpretation of the Dutch Book results given by Colin Howson and Alan Franklin (1994) (a related approach is given in Howson and Urbach 1989, ch. 3). They argue that an agent who has a certain degree of belief makes an implicit claim that certain betting odds are fair. On this assumption, an agent with incoherent degrees of belief is believing a pair of deductively inconsistent claims about fair betting odds. Howson and Franklin conclude that the probability axioms "are no more than (deductive) logic" (p. 457). But just as a particular degree of belief may, or may not, give rise to the ideally correlated betting preferences, a given degree of belief may or may not give rise to the correlated belief about fair betting odds. Even if we take degrees of belief to justify the correlated beliefs about fair bets, a degree of belief and a belief about betting are not the same thing. Once we see the possibility of this metaphysical connection being broken, it seems a mistake to hold that the real problem with incoherent degrees of belief lies in the claims about bets with which they are ideally correlated.

acknowledge that there is no necessary metaphysical connection between degrees of belief and bet evaluations. But one might hold that there are causal connections that hold in certain ideal situations, and that in those ideal situations violations of the probability calculus are always accompanied by preference inconsistencies. One might then point out, quite rightly, that finding norms for idealized situations is a standard and reasonable way of shedding light on normative aspects of situations where the idealizations do not hold.

But this, too, is unsatisfying. If the ultimate problem with incoherent degrees of belief lay just in their leading to preference inconsistencies, then there would seem to be no problem at all with incoherent beliefs in those non-ideal cases where they did not happen to give rise to inconsistent preferences. This seems quite unintuitive: there is something wrong with the beliefs of an agent who thinks P more likely to be true than (P ∨ Q), even if the psychological mechanisms that would ideally lead from these beliefs to the correlated preferences are for some reason disrupted. And it would involve quite a strain to suggest that the ultimate problem with such an agent's beliefs lay simply in the fact that these preferences would, in ideal circumstances, give rise to inconsistent preferences: there seems to be something wrong with thinking that P is more likely to be true than (P ∨ Q), quite apart from any effect this opinion might have on the agent's practical choices or preferences. Ultimately, to locate the problem with probabilistically incoherent degrees of belief in the believer's preferences, actual or counterfactual, is to mislocate the problem.

For these reasons, I think we must reject the pragmatic consistency interpretations of the DBAs. Should we, then, give up on the DBAs themselves? Perhaps not. It seems to me that the arguments have enough initial intuitive power that it would be disappointing, and even a bit surprising, if they turned out to be as thoroughly misguided as their pragmatic interpretations seem to make them. In the next section, I'll explore the possibility of making sense of the DBAs in a fully non-pragmatic way.

5.3 Dutch Book Arguments Depragmatized

Although the relationship between degrees of belief and the evaluations of betting odds to which they often give rise may not be as close as some have thought, there is, I think, a relationship that goes well beyond the rough psychological causal pattern. Putting aside any behaviorist or functionalist accounts of partial belief, it is initially quite plausible that, in ordinary circumstances, a degree of belief in P of, e.g., 2/3 that of certainty *sanctions as fair*—in one relatively pre-theoretic, intuitive sense—a monetary bet on P at 2 : 1 odds. Intuitively, the agent's level of confidence in P's truth provides *justification* for the agent's bet evaluation—it is part of what makes the bet evaluation a reasonable one.

Let us try to make the intuitive idea a bit more precise. To begin with, let us say that an agent's degree of belief in a certain proposition sanctions a bet as fair if it provides justification for evaluating the bet as fair—i.e. for being indifferent to taking either side of the bet. Clearly, this connection depends in any given case on the agent's values. If an agent values roast ducks more than boiled turnips, her belief that a coin is unbiased will not sanction as fair a bet in which she risks a roast duck for a chance of gaining a boiled turnip on the next coin flip. If she values the two equally, however, and values nothing else relevant in the context, she should be indifferent to taking either side of a bet, at one duck to one turnip, on the next flip of a coin she believes to be fair.

How does this general idea connect with monetary betting odds? It cannot, of course, be that any agent with 2/3 degree of belief in P is rationally obliged to agree to putting up $200 to the bookmaker's $100 on a bet the agent wins if P is true. Various factors may make it irrational for her to accept such bets. The value of money may be non-linear for her, so that, e.g., the 200th dollar would be worth less than the 17th. Or she may have non-monetary values—such as risk aversion—which affect the values she attaches to making the monetary bets. So, in general, we cannot correlate a person's degree of belief in P with the monetary odds at which it is reasonable for her to bet on P.

In order to sidestep these issues, let us concentrate for the time being on agents with value structures so simple that such considerations do not arise. Let us consider an agent who values money positively, in a linear way, so that the 200th dollar is worth exactly the same as the 17th. And let us suppose that he does not value anything else at all, positively or negatively. I'll call this sort of being a *simple agent*. For a simple agent, there does seem to be a clear relation between degrees of belief and the monetary odds at which it is reasonable for him to bet. If a simple agent has a degree of belief of, e.g., 2/3 that P, and if he is offered a bet in which he will win $1 if P is true and lose $2 if P is false, he should evaluate the bet as fair. The same would hold of a bet that would cost him $100 if P is true but would pay him $200 if P is false. I take these as very plausible normative judgments: any agent who values money positively and linearly, and who cares about nothing else, *should* evaluate bets in this way. This suggests the following principle relating a simple agent's degrees of belief to the bet evaluations it is reasonable for him to make.

> *Sanctioning*. A simple agent's degrees of belief sanction as fair monetary bets at odds matching his degrees of belief.[8]

Degrees of belief may in this way *sanction* certain bets as fair, even if the degrees of belief do not *consist in* propensities to bet, or even to evaluate bets, in the sanctioned way. The connection is neither causal nor definitional: it is purely normative.

Now one might wonder whether this normative claim begs the present question. After all, the matching between beliefs and betting odds is the same one that emerges from expected utility theory,

[8] "Matching" here is understood in the natural way, corresponding to the betting quotients mentioned in fn. 1 above. Thus, if one's degree of belief in proposition P is r, the matching odds would be $\$r{:}\,\$(1-r)$. If my degree of belief in P is 3/4, a bet I'd win if P were true, and in which I put up my 75¢ to my opponent's 25¢, would be at matching odds, as would a bet in which I put up $3 to my opponent's $1.

which already presupposes a probabilistic consistency requirement. But the intuitive normative connection between degrees of belief and bets need not derive from an understanding of expected utility theory; a person might see the intuitive relationship between bets and degrees of belief even if she could not begin to describe even roughly how the probability of P, Q, (P & Q), and (P ∨ Q) should in general relate to one another. Of course, there may be a sense in which our intuitions on these topics are all interrelated, and spring from some inchoate understanding of certain principles of belief and decision. But that seems unobjectionable; indeed, it is typical of situations in which we support a general formal reasoning theory by showing that it coheres with our more specific intuitions.[9]

Given this normative connection between an agent's degrees of belief and betting preferences, the rest of the DBA can be constructed in a fairly standard way. We may say that if a set of bets is logically guaranteed to leave an agent worse off, by his own lights, then there is something rationally defective about that set of bets. This general intuition may easily be applied to a simple agent in a straightforward way: since the simple agent cares solely and positively about money, a set of bets that is guaranteed to cost him money is guaranteed to leave him worse off, by his own lights. This yields the following principle.

> *Bet Defectiveness.* For a simple agent, a set of bets that is logically guaranteed to leave him monetarily worse off is rationally defective.

[9] It is also worth noting that even the "mitigated pragmatic consistency" interpretation of the DBA discussed above must presuppose a basic normative connection between degrees of belief and bet evaluations. On this view, degrees of belief lead causally to the correlated betting preferences in ideal circumstances. But one might ask: which circumstances are "ideal"? Why single out those circumstances in which degrees of belief lead to exactly the preferences that expected utility theory would dictate? The answer, it seems to me, is that we are intuitively committed to a certain normative relation between degrees of belief and preferences. Circumstances are "ideal" when, and because, this intuitively plausible relation obtains. If this answer is right, then what is perhaps the most controversial assumption in the non-pragmatic interpretation of Dutch Books given in the text also figures in the "mitigated pragmatic consistency" interpretation.

We now need a principle that connects the rational defectiveness in a set of bets to a rational defect in the degrees of belief that sanction those bets. But it is not generally true that, for any agent, a set of beliefs that sanctions each of a defective set of bets is itself defective. The reason for this stems from an obvious fact about values: in general, the values of things are dependent on the agent's circumstances. Right now, I would put quite a high value on obtaining a roast duck, but if I already had a roast duck in front of me, obtaining another would be much less attractive. This phenomenon applies to the prices and payoffs of bets as much as to anything else; thus there can be what one might call *value interference* effects between bets. The price or payoff of one bet may be such that it would alter the value of the price or payoff of a second bet. And this may happen in a way that makes the second unfair—even though it would have been perfectly fair, absent the first bet. Because of such value interference effects, it is not in general true that there is something wrong with an agent whose beliefs individually sanction bets that, if all taken together, would leave her worse off.

Of course, insofar as value interference effects are absent, the costs or payoffs from one bet will not affect the value of costs or payoffs from another. And if the values that make a bet worth taking are not affected by a given factor, then the acceptability of the bet should not depend on that factor's presence or absence. Thus in circumstances where value interference does not occur, bets that are individually acceptable should, intuitively, be acceptable in combination.

Fortunately, we already have before us a model situation in which value interference is absent: the case of the simple agent. The simple agent values money linearly; the millionth dollar is just as valuable as the first, and so the value of the costs and payoffs from one bet will not be diminished or augmented by costs or payoffs from another. Thus the following principle is, I think, quite plausible.

Belief Defectiveness. If a simple agent's beliefs sanction as fair each of a set of bets, and that set of bets is rationally defective, then the agent's beliefs are rationally defective.

It is worth noting that the intuitive appeal of Belief Defectiveness does flow, at least in part, from some general intuition about beliefs fitting together. So one might worry that the principle's intuitive plausibility presupposes a commitment to probabilistic coherence. Maher, criticizing a related principle, raises the following sort of worry. Consider a simple agent whose degree of belief in P is 1/3, yet whose degree of belief in not-P is also 1/3, violating probabilistic coherence. Such an agent's beliefs would sanction a defective set of bets.[10] But suppose one were to claim that the agent's beliefs were not themselves defective. We could not reply, without begging the question, by claiming that beliefs should fit together in the manner prescribed by the laws of probability.[11]

Nevertheless, this sort of example does not show that the plausibility of Belief Defectiveness is somehow intuitively dependent on the assumption of a probabilistic coherence requirement. The defect in the set of sanctioned bets lies in the way they fit together. The intuition behind Belief Defectiveness is that, absent value interference effects, this failure of the bets to fit together reflects a lack of fit between the beliefs that sanctioned those bets. But saying that the plausibility of the principle depends on a general intuition about beliefs fitting together does not mean that it depends intuitively on a prior acceptance of probabilistic coherence in particular. Belief Defectiveness would, I think, appeal intuitively to people who were quite agnostic on the question of whether, when A and B are mutually exclusive, the probability of $(A \vee B)$ was equal to the sum of the probability of A and the probability of B. The idea that beliefs should fit together *in that particular way* need not be embraced, or even understood, in order for a general fitting-together requirement along the lines embodied in our principle to be plausible. Thus while Belief Defectiveness is certainly contestable, it seems to me intuitively

[10] His degree of belief in P would sanction a bet costing $2 if P is true, and paying $1 if P is false. His degree of belief in not-P would sanction a bet costing $2 if not-P is true, and paying $1 if not-P is false. The set of two bets is guaranteed to cost him $1.

[11] Maher's point is in his (1997), in the section criticizing Christensen (1996).

plausible, quite independently of the conclusion the DBA is aiming to reach.

With the three more philosophical premises in place, all that is needed for a DBA is the mathematical part.

> *Dutch Book Theorem.* If an agent's degrees of belief violate the probability axioms, then there is a set of monetary bets, at odds matching those degrees of belief, that will logically guarantee the agent's monetary loss.

The argument proceeds as follows. Suppose a simple agent has probabilistically incoherent degrees of belief. By the Dutch Book Theorem, there is a set of monetary bets at odds matching his degrees of belief which logically guarantee his monetary loss. By Bet Defectiveness, this set of bets is rationally defective, and by sanctioning, each member of this set of bets is sanctioned by his degrees of belief. Then, by Belief Defectiveness, his beliefs are rationally defective. Thus we arrive at the following.

> *Simple Agent Probabilism.* If a simple agent's degrees of belief violate the probability axioms, they are rationally defective.

This distinctively non-pragmatic version of the DBA allows us to see why its force does not depend on the real possibility of being duped by clever bookies. It does not aim at showing that probabilistically incoherent degrees of belief are unwise to harbor for practical reasons. Nor does it locate the problem with probabilistically incoherent beliefs in some sort of preference inconsistency. Thus it does not need to identify, or define, degrees of belief by the ideally associated bet evaluations. Instead, this DBA aims to show that probabilistically incoherent beliefs are rationally defective by showing that, in certain particularly revealing circumstances, they would provide *justification* for bets that are rationally defective in a particularly obvious way. The fact that the diagnosis can be made *a priori* indicates that the defect is not one of fitting the beliefs with the way the world happens to be: it is a defect internal to the agent's belief system.

As set out above, the conclusion of the DBA has its scope restricted to simple agents. And this fact gives rise to a potentially troubling question: doesn't this deprive the argument of its interest? After all, it is clear that there are not, and have never been, any simple agents. What is the point, then, of showing that simple agents' beliefs ought to be probabilistically coherent?[12]

The answer to this question is that while the values of simple agents are peculiarly simple, the point of the DBA is not dependent on this peculiarity. The argument takes advantage of the fact that rational preferences for bets are informed jointly by an agent's values and an agent's representations of the world—her beliefs. In our thought-experiment, we consider how a certain set of beliefs would inform the betting preferences of an (imaginary) agent who cared only about one sort of thing, and cared about it in a very simple way (money is the traditional choice, but it's arbitrary; grains of sand would serve as well). This particularly transparent context allows us to see a clear intuitive connection between the set of beliefs and certain bets: given the simple values, the beliefs provide justification for evaluating the bets as fair. We show that, if the beliefs are incoherent, they would justify the imagined agent's preferring to take each of a set of bets that would logically guarantee his losing the only commodity he values. Given the agent's simple value structure, the problem with the set of bets cannot be that the costs or benefits of one bet affect the value of the costs or benefits of another. Rather, the problem is that there is no way the world could turn out that would make the set of bets work out well—or even neutrally—for the agent. In this sort of case, it seems to me that the overwhelmingly plausible diagnosis is that there is something intrinsically wrong with the representations of the world that justified the agent's preferences for these bets.

[12] This objection is similar to one considered by Kaplan, whose argument for a weakened version of probabilism incorporates the same assumptions about the agent's values. My answer is in part along lines roughly similar to Kaplan's (see Kaplan 1996, 43–4).

This is in part why it is important to be clear on the role that preferences play in the DBA. If the basic problem diagnosed in these cases were that the simple agent's preferences would get him into trouble, or even that the simple agent's preferences were themselves inconsistent, then one might well ask "Why is the correct conclusion that the degrees of belief are irrational *per se*, rather than that it is irrational to have incoherent beliefs if you are a simple agent?"[13] For if the basic defect were located in the simple agent's preferences, then it would be unclear why we should think that the problem would generalize to agents with very different preference structures. But the basic defect diagnosed in the simple agent is not a preference-defect. In severing the definitional or metaphysical ties between belief and preferences, the depragmatized DBA frees us from seeing the basic problem with incoherent beliefs as a pragmatic one, in any sense. Once the connection between beliefs and preferences is understood as normative rather than metaphysical, we can see that the simple agent's problematic preferences function in the DBA merely as a diagnostic device, a device that discloses a purely epistemic defect.

Thus, the lesson of the depragmatized DBA is not restricted to simple agents. Nor is it restricted to agents who actually have the preferences sanctioned by their beliefs. (In fact, the defect that, in simple agents, results in Dutch Book vulnerability may even occur in agents in whom no bet evaluations, and hence no bet evaluation inconsistencies, are present.) The power of the thought experiment depends on its being plausible that the epistemic defect we see so clearly when incoherent beliefs are placed in the value-context of the simple agent is also present in agents whose values are more complex. I think that this is quite plausible. There is no reason to think that the defect is somehow an artefact of the imagined agent's unusually simple value structure. So although an equally clear thought-experiment that did not involve simple agents might

[13] I owe this formulation of the question to an anonymous referee for *Philosophy of Science*.

have been more persuasive, the simple-agent-based example used in the depragmatized DBA above seems to me to provide powerful intuitive support for probabilism.[14]

5.4 Representation Theorem Arguments

If DBAs are the best-known ways of supporting probabilism, Representation Theorem Arguments (RTAs) are perhaps taken most seriously by committed probabilists.[15] RTAs approach an agent's beliefs and values in a more holistic way than do DBAs. The arguments begin by taking ideally rational preferences to be subject to certain intuitively attractive formal constraints, such as transitivity. They then proceed to demonstrate mathematically (via a Representation Theorem) that if an agent's preferences obey the formal constraints, they can be represented as resulting from a relatively unique[16] pair, consisting of a set of degrees of belief and a set of utilities, such that (1) the degrees of belief are probabilistically

[14] This point suggests another approach to the worry expressed in the text. If the monetary bets that figured in the simple-agent DBA were replaced by bets that paid off in "utiles" instead of dollars, the argument could be rewritten without the restriction to simple agents. (The idea here is not that the bets would be paid monetarily, with amounts determined by the monetary sums' utilities relative to the agent's *pre-bet* values; as Maher (1993, 97–8) points out, this would not solve the problem. The idea is that a bet on which an agent won, e.g., 2 utiles would pay her in commodities that would be worth 2 utiles at the time of payment. Because of value interference, a proper definition of the payoffs might have to preclude bets being paid off absolutely simultaneously, but I don't see this as presenting much of a problem.) Nevertheless, generalizing the DBA in terms of utiles would decrease the intuitive transparency of its premises. Insofar as the point of the argument is to provide intuitive support for probabilism, the more general argument would, I suspect, actually be less powerful.

[15] See e.g. Maher (1993, ch. 4.6) or Kaplan (1996, ch. 5).

[16] "Relatively" unique because, e.g., different choices of a zero point or unit for a utility scale might work equally well. Different representation theorems achieve different sorts of relative uniqueness. For present purposes, I'll put aside worries about the way particular versions of the RTA deal with failure of absolute uniqueness. Since the issues raised below would arise even if absolute uniqueness were achieved, I'll write as if the theorems achieved true uniqueness.

coherent, and (2) the preferences maximize expected utility relative to those beliefs and utilities. Thus, typical RTAs begin with some version of the following two principles.

> *Preference Consistency*. Ideally rational agents' preferences obey constraints C.

> *Representation Theorem*. If an agent's preferences obey constraints C, then they can be represented as resulting from some unique set of utilities U and probabilistically coherent degrees of belief B relative to which they maximize expected utility.

Clearly, these principles alone are not enough to support the intended conclusion. The fact that an agent's preferences can be represented as resulting from some U and B does not show that U and B are that agent's actual utilities and degrees of belief. Typically, RTA proponents rely in their arguments on some principle positing a tight definitional or constitutive connection between an agent's preferences and her beliefs and utilities. The precise form of the principle making the connection may vary, and it may receive little philosophical comment, but the following sort of connection is taken to emerge from the argument.

> *Representation Accuracy*. If an agent's preferences can be represented as resulting from unique utilities U and probabilistically coherent degrees of belief B relative to which they maximize expected utility, then the agent's actual utilities are U and her actual degrees of belief are B.

Given these three principles, we get:

> *Probabilism*. Ideally rational agents have probabilistically coherent degrees of belief.

Thus understood, representation theorems provide for a particularly interesting kind of argument. From a normative constraint on preferences alone, along with some mathematics and a principle about the accuracy of certain representations, we can derive a normative constraint on degrees of belief.

The mathematical meat of this argument—the Representation Theorem itself—has naturally received most of the attention. Of the more purely philosophical principles, Preference Consistency has been discussed much more widely. Some claim that its constraints on preferences are not satisfied by real people—and, more interestingly, that violations of the constraints are not irrational. I'll pass over this discussion for the present, assuming that the constraints are plausible rational requirements.[17] Instead, as with the DBA, I'll focus on the purported connection between the clearly epistemic and the pragmatic aspects of rationality, as summarized in the Representation Accuracy Principle. Suppose that an agent has preferences that would accord with expected utility (EU) maximization relative to some unique U and B. Why should we then take U and B to be her actual utilities and—most importantly for our purposes—beliefs?[18]

Representation Accuracy posits that a particular connection holds among agents' preferences, utilities, and beliefs. That there is, in general, some connection of very roughly the sort posited is an obvious truism of folk psychology. People do typically have preferences for options based on how likely they believe the options are to lead to outcomes they value, and on how highly they value the possible outcomes. But the cogency of the RTA requires a connection much tighter than this.

We can start to see why by noting that the purposes of the RTA would not be served by taking Representation Accuracy as a mere empirical regularity, no matter how well confirmed. For the purported empirical fact—that having probabilistically coherent beliefs is, given human psychology, causally necessary for having

[17] Patrick Maher (1993) provides very nice explanations of—and defenses against—these objections.

[18] Lyle Zynda (2000) focuses on this aspect of the RTA; he calls it "The Reality Condition." My overall sketch of the RTA is very similar to Zynda's, though my conclusions diverge quite widely from his.

consistent preferences—would at best show probabilistic coherence valuable in a derivative and contingent way. After all, one might discover empirically that, given human psychology, only those whose beliefs were unrealistically simple, or only those suffering from paranoid delusions, had preferences consistent enough to obey the relevant constraints. If a representation theorem is to provide a satisfying justification for Probabilism—if it is to show that the rules of probability provide a correct way of applying logic to degrees of belief—then the connection between preferences and beliefs will have to be a deeper one.

In fact, RTA proponents do posit deeper connections between preferences and beliefs. Like DBA proponents, they typically take degrees of belief (and utilities) to be in some sense *defined* by preferences. Taken unsympathetically, this suggests some sort of operationalism or related notion of definition via analytic meaning postulates. But it seems to me that a more charitable reading of the argument is available.

Let us begin with a look at the role that degrees of confidence play in psychological explanation. Clearly, we often explain behavior—especially in deliberate choice situations—by invoking degrees of confidence. Often, these explanations seem to proceed via just the sort of principle that lies behind Representation Accuracy. We explain someone's selling a stock by an increase in his confidence that it will soon go down, assuming that his choice is produced by his preferences, which themselves result from his beliefs and utilities in something like an EU-maximizing way.

Thus, we might see Representation Accuracy as supported by the following kind of thought: "The belief-desire model is central to the project of explaining human behavior. Degrees of belief are posited as working with utilities to produce preferences (and hence choice-behavior). The law connecting beliefs and utilities to preferences is that of maximizing EU. So beliefs are, essentially, that which, when combined with utilities, determine preferences via EU-maximization." Patrick Maher, in a sophisticated recent defense of the RTA, writes:

I suggest that we understand probability and utility as essentially a device for interpreting a person's preferences. On this view, an attribution of probabilities and utilities is correct just in case it is part of an overall interpretation of the person's preferences that makes sufficiently good sense of them and better sense than any competing interpretation does.... [I]f a person's preferences all maximize expected utility relative to some p and u, then it provides a perfect interpretation of the person's preferences to say that p and u are the person's probability and utility functions. (Maher 1993, 9)

This approach toward defining degrees of belief by preferences need not be fleshed out by any naive commitment to operationalism, or to seeing the relevant definition as analytic or a priori. And the definition need not be the simple sort that figures in some presentations of the DBA, where an agent's degree of belief is defined in terms of very particular betting preferences. We needn't even see the agent's preferences as epistemically privileged, compared with her beliefs and utilities. Jeffrey writes:

In fact, I do not regard the notion of preference as epistemologically prior to the notions of probability and utility. In many cases we or the agent may be fairly clear about the probabilities the agent ascribes to certain propositions without having much idea of their preference ranking, which we thereupon deduce indirectly, in part by using probability considerations. The notions of preference, probability, and utility are intimately related; and the object of the present theory is to reveal their interconnections, not to "reduce" two of them to one of the others. (Jeffrey 1965*b*, 220–1)

The envisioned account of graded belief might thus be understood as a more holistic scientific definition, combining elements of conceptual refinement with empirical investigation. Beliefs turn out to be something like functional or dispositional properties of people, defined, along with utilities, by their causal connections to the agent's utilities, other beliefs, and preferences. On such a view, the fact that a strong belief that a stock will go down produces a strong preference to sell it is neither an analytic truth nor a mere empirical regularity. But part of what *constitutes* a given agent's having a strong belief that the stock will go down is precisely her disposition (given the usual utilities) to prefer selling the stock. Thus there is a metaphysical or constitutive connection among degrees of belief, utilities, and preferences. This idea has obvious connections to functionalist theories in mainstream philosophy of mind.

Nevertheless, this claim about the nature of beliefs cannot represent mere naked stipulation. If the definition is to have relevance to epistemology, the entities it defines must be the ones we started wondering about when we began to inquire into rational constraints on belief. And it seems to me that there are grounds for doubting that the envisaged definition will pass this test.

One worry we might have on this score is that the EU-based definition offered by RTA proponents is not the only one that would fit the somewhat vague intuitions we have about, e.g., the stock-selling case. Suppose we have an agent whose preferences fit the constraints and can thus be represented as resulting from coherent beliefs B and utilities U. Zynda argues that there will be another belief-function, B′, which is probabilistically incoherent, yet which may be combined with U (non-standardly) to yield a valuation function fitting the agent's preference ordering equally well.[19] Zynda concludes that the RTA can be maintained, but that we must justify our choice of B over B′. Endorsing Maher's view that probabilities and utilities are "essentially a device for interpreting a person's preferences," he favors taking a less-than-fully realistic view of beliefs, on which our choice of B over B′ can be made on frankly pragmatic grounds.

It seems to me, however, that the RTA proponent faces complexities beyond those revealed by Zynda's example. For our question is not merely whether the proposed definition uniquely satisfies our intuitions about deliberate choice cases. We want to know how closely this definition fits our intuitive concept in general. Let us

[19] Zynda's B′ is a linear transformation of B; the non-standard valuation function is tailored to compensate for this transformation; see (Zynda 2000, 8 ff.).

look, then, a bit more broadly at the pre-(decision-)theoretic notion of strength of belief.

To begin with, it is obvious that anyone can tell by quick introspection that she is more confident that the sun will rise tomorrow than that it will rain tomorrow. But it is not at all clear that this aspect of our common notion jibes with the envisioned definition. And, in fact, some RTA proponents have considered this sort of worry. Ramsey, dubious of measuring degrees of belief by intensity of introspected feeling, saw his definition as capturing "belief qua basis of action," arguing that even if belief-feelings could be quantified, beliefs as bases of action were what was really important (1926, 171–2). Ellery Eells (1982, 41–3) also supports seeing beliefs as dispositions to action by developing Ramsey's criticism of measuring degrees of belief via feelings of conviction.

This discounting of the introspective aspect of our pre-theoretic notion is not an unreasonable sort of move to make. If a common concept is connected both to quick identification criteria and to deeper explanatory concerns, we do often override parts of common practice. Thus, we might discount introspectively based claims about degrees of belief if and when they conflict with the criteria flowing from our explanatory theory. This move is made more reasonable by the fact, emphasized by some RTA proponents, that our introspective access seems pretty vague and prone to confusion.

But the general worry—that the preference-based definition leaves out important parts of our pre-theoretic notion—is not this easily put aside. For one thing, it seems clear that, even within the realm of explaining behavior, degrees of belief function in ways additional to explaining preferences (and thereby choice-behavior). For example, we may explain someone coming off well socially on the basis of her high confidence that she will be liked. Or we may explain an athlete's poor performance by citing his low confidence that he will succeed.

Examples like this can be multiplied without effort. And it does not seem that anything involving choice between options, or, really, any aspect of preferences, is being explained in such cases. Rather, it is an important psychological fact that a person's beliefs—the way she represents the world—affect her behavior in countless ways that have nothing directly to do with the decision theorist's paradigm of cost–benefit calculation.

Moreover, degrees of belief help explain much more than behavior. We constantly invoke them in explanations of other psychological states and processes. Inference is one obvious sort of case: we explain the meteorologist's increasing confidence in rain tomorrow by reference to changes in her beliefs about the locations of weather systems. But beliefs are also universally invoked in explanations of psychological states other than beliefs (and other than preferences). We attribute our friend's sadness to her low confidence in getting the job she's applied for. We explain a movie character's increasing levels of fear on the basis of his increasing levels of confidence that there is a stranger walking around in his house. The connections between beliefs and other psychological states invoked in such explanations are, I think, as basic, universal, and obvious as the central connections between beliefs and preferences that help explain behavior.

Beliefs may also have less obvious non-behavioral effects. Every reputable drug study controls for the placebo effect. According to received wisdom, people's confidence that they are taking effective medicine reliably causes their conditions to improve, often in physiologically measurable ways. The exact mechanisms behind the placebo effect are unclear (and one recent study suggests that this effect is far less prevalent than it is standardly taken to be).[20] But insofar as the placebo effect is real, it is not explained by any disposition of the patients to have preferences or make choices that maximize utility relative to a high probability of their having taken effective medicine.

[20] See Hrobjartsson and Gotzsche (2001). As might be expected, the study's conclusions are somewhat controversial. The authors conclude that there is no justification for using placebos therapeutically, but they do not recommend the elimination of placebos in clinical trials.

Thus, it turns out that the RTA proponents' problem with accommodating introspective access to our degrees of belief represents the tip of a very large iceberg. True, degrees of belief are intimately connected with preferences and choice-behavior. But they are also massively and intimately connected with all sorts of other aspects of our psychology (and perhaps even physiology). This being so, the move of settling on just one of these connections—even an important one—as definitional comes to look highly suspicious.

This is not to deny that beliefs may, in the end, be constituted by their relations to behaviors and other mental states—by their functional role in the agent. But even functionalists have not limited their belief-defining functional relations to those involving preferences, and it is hard to see any independent motivation for doing so. And if the preference-explaining dispositions are only parts of a much larger cluster of dispositions that help to constitute degrees of belief, then it is hard to see how Representation Accuracy, or Maher's claim quoted above, can be maintained. After all, a given interpretation of an agent's degrees of belief might maximize expected-utility fit with the agent's preferences, while a different interpretation might fit much better with other psychological–explanatory principles. In such cases of conflict, where no interpretation makes all the connections come out ideally, there is no guarantee that the best interpretation will be the one on which the agent's preferences accord perfectly with maximizing EU. And if it is not, then even an agent whose preferences obey Preference Consistency may fail to have probabilistically coherent degrees of belief. Thus it seems that even if we take a broadly functionalist account of degrees of belief, Representation Accuracy is implausible.

Moreover, it is worth pointing out that the assumption that beliefs reduce to dispositional or functional states of any sort is highly questionable. The assumption is clearly not needed in order to hold, e.g., that preferences give us a quite reliable way of measuring degrees of belief, or that beliefs play a pervasive role in explaining preferences and other mental states and behaviors. Beliefs can enter into all sorts of psychological laws, and be known through these laws, without being reductively defined by those laws. They may, in short, be treated as typical theoretical entities, as conceived of in realistic philosophy of science.[21] If the connections between beliefs and preferences have the status of empirical regularities rather than definitions—if the connections are merely causal and not constitutive—then the RTA would fail in the manner described above. It would be reduced to showing that, given human psychology (and probably subject to extensive ceteris paribus conditions), coherent beliefs do produce rational preferences. This is a long way from showing that coherence is the correct logical standard for degrees of belief.

In retrospect, perhaps it is not surprising that the ironclad belief–preference connection posited in Representation Accuracy fails to be groundable in—or even to cohere with—a plausible metaphysics of belief. Degrees of belief are not merely part of a "device for interpreting a person's preferences." Beliefs are our way of representing the world. They come in degrees because our evidence about the world justifies varying degrees of confidence in the truth of various propositions about the world. True, these representations are extremely useful in practical decisions; but that does not reduce them to mere propensities to decide. After all, it seems perfectly coherent that a being could use evidence to represent the world in a graded manner without having utilities or preferences at all!

Such a being would not be an ordinary human, of course. But even among humans, we can observe differences in apparent preference intensities. (Clearly, intersubjective comparisons are difficult, but that hardly shows that intersubjective differences are unreal.) I don't think that we would be tempted to say, of a person affected with an extreme form of diminished affect—a person who

[21] For an argument showing that functionalist accounts of mental states are fundamentally incompatible with a robust kind of scientific realism, see Derk Pereboom (1991).

had no preferences—that he had no beliefs about anything. After all, it is obvious from one's own case that one cares about some things much more than one cares about others. One can easily imagine one's self coming to care less and less about more and more things. But insofar as one can imagine this process continuing to the limit, it does not in the slightest seem as if one would thereby lose all beliefs.

One might object that a preferenceless being would still have *dispositions* to form EU-maximizing preferences, in circumstances where it acquired utilities. But what reason would we have to insist on this? Given the being's psychological makeup, it might be impossible for it to form utilities. Or the circumstances in which it would form utilities might be ones where its representations of the world would be destroyed or radically altered.

The suggestion that having a certain degree of belief reduces to nothing more than the disposition to form preferences in a certain way should have struck us as overly simplistic from the beginning. After all, it is part of common-sense psychology that, e.g., the strength of an agent's disposition to prefer bets on the presence of an intruder in the house will be strongly correlated with the strength of the agent's disposition to feel afraid, and with the strength of his disposition to express confidence that there's an intruder in the house, etc. The view that identifies the belief with just one of these dispositions leaves the other dispositions, and all the correlations among them, completely mysterious. Why, for example, would the brute disposition to form preferences in a certain way correlate with feelings of fear?[22]

This point also makes clear why it won't do to brush the problem aside by claiming only to be discussing a particular sort of belief, such as "beliefs qua basis of action." It is not as if we have one sort of psychological state whose purpose is to inform preferences, and a separate sort of state whose purpose is to guide our emotional lives, etc. As Kaplan notes (in arguing for a different point), "You have

[22] Sin yee Chan (1999) makes a parallel point about emotional states.

only one state of opinion to adopt—not one for epistemic purposes and another for non-epistemic purposes" (1996, 40). What explains the correlations is that they all involve a common psychological entity: the degree of belief.

Degrees of belief, then, are psychological states that interact with utilities and preferences, as well as with other aspects of our psychology, and perhaps physiology, in complex ways, one of which typically roughly approximates EU-maximization. Whether we see the connection between the preference-dispositions and beliefs as partially constitutive (as functionalism would) or as resulting from purely contingent psychological laws (as a more robust realism might) is not crucial here. For neither one of these more reasonable metaphysical views of belief can support Representation Accuracy. If this is correct, then it becomes unclear how a Representation Theorem, even in conjunction with Preference Consistency, can lend support to Probabilism.[23]

5.5 De-metaphysicized Representation Theorem Arguments

Representation Accuracy asserted that whenever *any* agent's preferences maximized EU relative to a unique U and B, the agent's actual utilities and beliefs were U and B. The suspicious metaphysics was needed to ensure the universality of the posited preference–belief connection. But the RTA's conclusion does not apply to all

[23] Brad Armendt (1993) notes that in both the DBA and the RTA the connections between beliefs and preferences may be challenged. But he holds that the move of defining beliefs in terms of preferences is inessential. The RTA's assumption about the belief–preference connection applies in "uncomplicated cases where EU is most appropriate" (1993, 16). This point of Armendt's seems correct. But acknowledging that the belief–preference connection actually holds only in certain cases threatens to undermine the RTA. We are left needing a reason for thinking that the situations in which the belief–preference connection does hold are normatively privileged. Otherwise, it is hard to see why a result that applies to these cases—that Preference Consistency requires probabilistic consistency—would have any general normative significance. The next section attempts to provide just such a reason.

agents—only to ideally rational ones. Thus, the purpose of the RTA could be served without commitment to the preference–belief connection holding universally; it would be served if such a connection could be said to hold for all *ideally rational* agents.

Now one might well be pessimistic here—after all, if agents in general may have degrees of belief that do not match up with their utilities and preferences in an EU-maximizing way, why should this be impossible for ideally rational agents? The answer would have to be that the EU-maximizing connection is guaranteed by some aspect of ideal rationality. In other words, the source of the guarantee would be in a *normative*, rather than a metaphysical, principle.

This basic idea is parallel to the one exploited in the depragmatized DBA: to substitute a normative connection for a definitional or metaphysical one. In the RTA, we already assume that an ideally rational agent's preferences are consistent with one another in the ways presupposed in the obviously normative Preference Consistency principle. The present proposal is that, in addition, an ideally rational agent's preferences must cohere in a certain way with her beliefs. Of course, we cannot simply posit that such an agent's preferences maximize EU relative to her beliefs and utilities. Expected utility is standardly defined relative to a probabilistically coherent belief function. So understood, our posit would blatantly beg the question: if we presuppose that ideal rationality requires maximizing EU in this sense, then the rest of the RTA, including the RT itself, is rendered superfluous. Nevertheless, I think that a more promising approach may be found along roughly these lines.

Let us begin by examining the basic preference–belief connection assumed to hold by RTA proponents such as Savage (1954) and Maher. As noted above, Representation Accuracy emerges from a more specific belief–preference connection made in the course of the RTA. In proving their results, Savage and Maher first define a "qualitative probability" relation. This definition is in terms of preferences; it is at this point that the connection between preferences and beliefs is forged. The arguments then go on to show how (under specified conditions) a unique quantitative probability function corresponds to the defined qualitative relation. Maher explains the key definition of qualitative probabilities intuitively as follows:

> We can say that event B is more probable for you than event A, just in case you prefer the option of getting a desirable prize if B obtains, to the option of getting the same prize if A obtains.[24] (Maher 1993, 192)

Now it seems to me that there is something undeniably attractive about the idea that, in general, when people are offered gambles for desirable prizes, they will prefer the gambles in which the prizes are contingent on more probable propositions. However, in light of the arguments above, we should not follow Savage and Maher in taking this sort of preference–belief correspondence to *define* degrees of belief. In fact, we should not even assume that the connection holds *true* for all agents (or even for all agents whose preferences satisfy the RTA's constraints on preferences). Instead, we may take this sort of preference–belief connection to be a normative one, which holds for all ideally rational agents.

Seen as a claim about the way preferences *should* connect with beliefs, the connection posited in the RTA would amount to something like the following.

> *Informed Preference.* An *ideally rational* agent prefers the option of getting a desirable prize if B obtains to the option of getting the same prize if A obtains, just in case B is more probable for that agent than A.[25]

This normative principle avoids the universal metaphysical commitments entailed by the definitional approach. We may maintain such a principle while acknowledging the psychological possibility of a certain amount of dissonance between an agent's degrees of belief and her preferences, even when those preferences are

[24] The formal definition which cashes out this intuitive description is quite complex, and is premised on the agent's preferences satisfying certain conditions.

[25] This is, of course, an informal statement. Like Maher's informal definition above, it must be understood as applying only when certain conditions are met.

consistent with one another. At the same time, the principle forges the preference–belief connection for all ideally rational agents, who are anyway the only ones subject to the RTA's desired conclusion.[26]

Suppose, then, that the RTA was formulated using a suitably precise version of Informed Preference. Of course, this sort of RTA would not support the principle of Representation Accuracy—but, as we have seen, this is as it should be. What would emerge from the reformulated RTA would be Representation Accuracy's normative analogue.

> *Representation Rationality.* If an *ideally rational* agent's preferences can be represented as resulting from unique utilities U and probabilistically coherent degrees of belief B relative to which they maximize expected utility, then the agent's actual utilities are U and her actual degrees of belief are B.

This principle, no less than the rejected Representation Accuracy, may be combined with Preference Consistency and a Representation Theorem to yield Probabilism.

The RTA thus understood would presuppose explicitly a frankly normative connection between beliefs and preferences, something the RTA as standardly propounded does not do. Such an argument will thus need to be in one way more modest than the metaphysic-

[26] A principle much like Informed Preference is endorsed by Kaplan, in the course of giving his decision-theoretic argument for a weakened version of Probabilism which Kaplan terms "Modest Probabilism": "you should want to conform to the following principle.
Confidence. For any hypotheses P and Q, you are more confident that P than you are that Q *if and only if* you prefer ($1 if P, $0 if ~P) to ($1 if Q, $0 if ~Q)" (1996, 8).
Kaplan presents Confidence not as a definition, but as a principle to which we are committed (under suitable conditions) by reason. Kaplan's book is not concerned primarily with the issues we've been concentrating on: he is concerned to present an alternative to the Savage-style RTA which is much simpler to grasp, and which yields a weaker constraint on degrees of belief, a constraint that avoids certain consequences of Probabilism which Kaplan finds implausible. But while Kaplan does not discuss his departure from Savage's definitional approach to the connection between preferences and degrees of belief, his argument for Modest Probabilism exemplifies the general approach to RTA-type arguments advocated here.

ally interpreted RTA: it cannot purport to derive normative conditions on beliefs in a way whose only normative assumptions involve conditions on preferences alone.

Still, strengthening the RTA's normative assumptions in this way does not render it question-begging, as simply assuming EU maximization would have. The intuitive appeal of Informed Preference—which forges the basic belief–preference connection, and from which Representation Rationality ultimately derives—does not presuppose any explicit understanding of the principles of probabilistic coherence. The principle would, I think, appeal on a common-sense level to many who do not understand EU, and who are completely unaware of, e.g., the additive law for probabilities.

Thus understood, the RTA still provides an interesting and powerful result. From intuitively appealing normative conditions on preferences alone, along with an appealing normative principle connecting preferences with beliefs, we may derive a substantial normative constraint on beliefs—a constraint that is not obviously implicit in our normative starting points. The argument is also freed from its traditional entanglement with behaviorist definition or other fishy metaphysics. Moreover, this frankly normative approach to the RTA answers the question posed above: how would a result that held in only special situations support a general normative requirement? On the approach advocated here, since the posited preference–belief connection is justificatory rather than causal or constitutive, we need not suppose that it ever holds exactly, even in uncomplicated cases. Thus, it seems to me that the RTA may be de-metaphysicized successfully; once this is done, the argument can lend substantial support to Probabilism.

5.6 Preferences and Logic

Both the RTA and the DBA attempt to support probabilism by exploiting connections between an agent's degrees of belief and her preferences. Both arguments have traditionally been tied to assumptions that try to secure the belief–preference connections by definitional or metaphysical means. But the metaphysically intimate connections between beliefs and preferences that have been posited by proponents of preference-based arguments for probabilism sit uneasily with our pre-theoretic understanding of what belief is. This tension is surely part of what is expressed when Ramsey restricts his interest to "beliefs qua basis for action," or when Jeffrey acknowledges that our pre-theoretic notion of belief is "only vestigially present in the notion of degree of belief." It is understandable that many epistemologists have been reluctant to embrace arguments that treat belief as part of a "device for interpreting a person's preferences."

A related point concerns the status of logical norms for graded belief. Standard logical properties of propositions, and relations among them, may be used to constrain rational graded belief via the probability calculus. This is not an unnatural suggestion. But it is unnatural to suppose that the illogicality or lapse of epistemic rationality embodied in incoherent graded beliefs is, at bottom, a defect in the believer's (actual or counterfactual) preferences. Any argument that locates the irrationality of probabilistically incoherent graded belief in the believer's preferences invites the suspicion that it is addressed to pragmatic, not epistemic, rationality. It makes it seem that probabilism is doing something quite different from what deductive cogency conditions were supposed to do for belief on the traditional binary conception.

We've seen that the definitional or metaphysical connections traditionally posited to underpin the preference-based arguments must be discarded. Fortunately, this need not mean discarding the insights that lie at the bottom of the RTA and DBA. For in each case, the argument's insights can be prised apart from the unsupportable assumptions. In each case, the insights can be preserved by seeing the belief–preference connections as straightforwardly normative rather than metaphysical. Once this is done, we see that the arguments apply to beliefs that are no more essentially pragmatic than binary beliefs have traditionally been thought to be.

On this interpretation, probabilism is nothing more than a way of imposing traditional logic on belief—it's just that this turns out to require that belief be seen in a more fine-grained way than it often has been. When we see belief as coming in degrees, and see logic as governing the degree to which we believe things, rather than as governing some all-or-nothing attitude of acceptance, probability theory is the overwhelmingly natural choice for applying logic to belief. The preference-based arguments supply natural support for this choice.

The best way of looking at both arguments is as using connections between beliefs and preferences purely diagnostically: in neither case should we see the argument as showing that the defect in incoherent beliefs really lies in the affected agent's preferences. Nor should we even see the problem as consisting in the beliefs' failure to *accord* with rational preferences. Beliefs are, after all, more than just a basis of action. The defect inherent in beliefs that violate probabilism should be seen as primarily epistemic rather than pragmatic. The epistemic defect shows itself in pragmatic ways, for a fairly simple reason: The normative principles governing preferences must of course take account of the agent's information about how the world is. When the agent's beliefs—which represent that information—are intrinsically defective, the preferences informed by those defective beliefs show themselves intrinsically defective too. But in both cases, the preference defects are symptomatic, not constitutive, of the purely epistemic ones.

Though the two preference-based arguments are similar, there are some interesting differences between them. The RTA's Informed Preference principle is simpler than the DBA's Sanctioning. The RTA also applies directly to any rational agent. But the RTA depends on some fairly refined claims about conditions on rational preferences, claims that some have found implausible. The DBA, though it applies directly only to simple agents, does not require taking the RTA's Preference Consistency principles as premises.

I suspect that different people will quite reasonably be moved to different degrees by these two arguments; and I don't see much point in trying to form very precise judgments about the arguments' relative merits. Neither one comes close to being a knockdown argument for probabilism, and non-probabilists will find contestable assumptions in both. But each of these arguments, I think, provides probabilism with interesting and non-question-begging intuitive support. Each shows that probabilism fits well with (relatively) pre-theoretic intuitions about rationality. And that may be the best one can hope for, in thinking about our most basic epistemic principles.

6 LOGIC AND IDEALIZATION

EVEN if the arguments we've been considering make an attractive case, in the abstract, for taking logic to constrain ideally rational belief by way of the probability calculus, one might worry that the whole approach of modeling rational belief in this formal way is unsound, because it embodies such a high degree of idealization. This sort of worry arises in different ways, some of which apply particularly to probabilistic models, and others of which would also apply to models based on deductive cogency. In the first category are worries prompted by the following sort of observation: although it seems clear that we believe some things and not others, and even that we believe some things more strongly than others, it does not at all seem as if we have the sort of numerically precise degrees of belief that figure in probabilistic coherence. In the second category are worries prompted by the observation that logical perfection is far beyond any human's capacity to achieve. In each case, the worry is that excessive idealization vitiates the normative significance of the formal model. Let us consider these worries in turn.

6.1 Vague Beliefs and Precise Probabilities

Suppose that one is convinced that, insofar as an agent's beliefs are representable by numerical degrees, those beliefs should obey the laws of probability. One might yet doubt that these laws could apply in any straightforward way to the sorts of graded beliefs agents really have. A common reason for skepticism on this point derives from worries about the logic's basic representation of graded beliefs by precise numbers. These worries are easy to generate; all it takes, as I. J. Good (1962, 81) noted, is "the sarcastic request for an estimate correct to twenty decimal places, say, for the probability that the Republicans will win the election." In fact, if one considers almost any ordinary proposition, one will be at a loss to say with any great degree of precision what one's degree of belief in that proposition is.

Now the fact that one cannot introspectively determine one's own degrees of belief to twenty decimal places surely does not show that one does not have such precise degrees of belief. But the point is not just about introspection. I take it as highly plausible that, in the vast majority of cases, people don't have degrees of belief that are precise to anything remotely like twenty decimal places. If beliefs were inscribed in an unambiguously precise language of thought in our heads, and if they were written in a notation with the gradation between white and black constituting degrees of credence from one to zero, then perhaps we would, after all, have numerically very precise degrees of belief (although even in this case, it's doubtful that degrees of belief would be precise to twenty decimal places). But on a more realistic account—one, for example, on which beliefs are realized in unimaginably complex neural configurations, and constituted in a way that relates them to other psychological phenomena such as preferences, emotions, inferential tendencies, and verbal and non-verbal behavioral dispositions—it would be surprising if there turned out to be such a thing as a real person's precise degree of belief in almost any proposition, even to two decimal places. Beliefs, like economic recoveries or other complexly constituted entities, will be vague.[1]

Now this point in itself, I think, should not be so worrisome. A formal model of rational belief may legitimately be idealized in

[1] Of course, I do not mean to suggest that those who model rational beliefs probabilistically ever do represent real people's beliefs by using 20—or even five—significant digits. The worry we are considering is that the probabilistic model of belief, as stated in the abstract, represents degrees of belief by real numbers, and real numbers are, by their very nature, absolutely precise.

more than one dimension. As has been noted above, such a model may idealize normatively: it may seek to represent ideally rational beliefs—beliefs that exhibit a kind of perfection of which actual humans are incapable. But it may also idealize in the way in which countless purely descriptive models idealize: it may assign a number to a quantity whose application to real instances is not completely precise.

It's worth remembering in this context that most of our common-sense and even scientific quantitative concepts, when applied to actual objects, involve the same sort of idealization. Consider the mean radius (in kilometers) of a planet; the volume (in cubic meters) of a lake; the concentration (in parts per million) of a hormone in an animal's blood; the height (in meters) of a tree; or the pH of a chemical sample. How many such quantities are correctly taken to be precise to the twentieth decimal place? I would conjecture that very few are. And of those that are, the same sort of question will arise at the 200th decimal place, or the 2,000th—this despite the fact that the models we use to understand the principles governing these entities use real numbers to represent the relevant quantities, and real numbers are fully precise.

It's a commonplace observation that when one applies a mathematical model to a real-world situation, the particular values one employs are only approximate, due to the inevitable inaccuracies of measurement. Clearly, this does not vitiate the interest or usefulness of the model; indeed, some of the basic skills learned in introductory science courses involve handling numerical models in a way that is sensitive to measurement inaccuracies (for example by using notational conventions for numbers that encode information about level of accuracy). But the fact that the quantities discussed above cannot correctly be thought to be precise to twenty decimal places is not just due to measurement inaccuracy. Consider, for example, the fact that the borders of planets, lakes, blood systems, trees, and chemical samples are vague at the microscopic level (and often at the macroscopic level). This sort of vagueness will not typically be reflected in the numbers we use to model the properties of these objects. But just as the inaccuracy of our measurements does not vitiate the interest of our formal models, neither does the fact that the quantities we measure are themselves vague. Clearly, our formal models routinely idealize away this sort of vagueness while remaining perfectly useful and even deeply illuminating. Indeed, in many areas of knowledge, models that employ this sort of idealization are the only kind we have ever had, or will ever have.

Thus it would be a big mistake to reject probabilistic models of rational belief merely on the ground that they fail the "twenty decimal place" test. As we saw above, there's clearly a real phenomenon of beliefs coming in degrees. And rationality clearly puts constraints on these degrees. Insofar as we are trying to model the way these degrees of belief should ideally fit together, representing these degrees by numbers seems entirely reasonable, although we realize from the beginning that our formal model (like so many others) embodies numerical precision, while the phenomenon being modeled is vague. So the descriptive claim that our actual degrees of belief are vague should not in itself undermine the project of using probability logic to characterize rules of rationality for graded belief.

But it would also be a mistake to think that the above discussion covered all of the worries one might have about the probabilistic model's use of precise numbers to represent vague degrees of belief. Another sort of worry flows from the explicitly normative aspirations of the probabilistic model.

We might start by noticing that in some cases where an agent fails to have a precise degree of belief, it is due to rational failure on the agent's part. Suppose, for example, that my degree of belief that the die will show 2 is 1/6, and my degree of belief that it will show 3 is also 1/6, yet my degree of belief in the proposition that it will show either 2 or 3 is vague, because I haven't thought the matter through. In such a case, it seems entirely appropriate for a normative theory to specify that an ideally rational agent with the first two degrees of belief would have 1/3 degree of belief in the disjunction.

Such cases of mismatch between our model and actual agents' credences are simply due to the fact that our model is idealized along the normative dimension.

However, the die example is atypical. In many ordinary cases where a person seems not to have a precise degree of belief in a proposition, we cannot see any particular degree of belief that he obviously ought to have. Indeed, it may well be claimed that for many propositions, in many evidential situations, it is just not the case that there is some unique precise degree of belief that it would be rational to have. This claim seems particularly plausible if we do not try to consider the proposition in question while holding all of the agent's other degrees of belief fixed. Suppose instead that we ask the following sort of question: given a particular evidential situation, conceived of as something other than an agent's whole set of degrees of belief, is there a unique probability function—i.e. a unique assignment of precise degrees of belief to each proposition—that would be ideally rational in that situation? I think that it is at least somewhat plausible that the answer to this question is "no."[2]

If this is right, it raises a normatively based question about the relationship between rational belief and the rules of probability. If there is no precise degree of belief that is required by rationality in a particular evidential situation, shouldn't our idealized model then include a way of representing epistemic attitudes other than precise degrees of belief?

The answer to that question will depend in part on how one sees the relationship between evidential situations and rationality. Perhaps the simplest way of seeing the relationship is as follows:

[2] I've put the question vaguely, in terms of "evidential situations." I intend this to be as neutral as possible among different conceptions of what it is that constrains rational belief evidentially. Some would take, e.g., perceptual experiences and apparent memories to be important here. Others would take a certain class of evidential beliefs (which may be required to have probability 1 themselves). On some global pure coherence views, there may be no distinction between an agent's evidential situation and her whole set of beliefs. But among those who recognize the distinction, I suspect that many would hold that the sorts of evidential situations in which we typically find ourselves do not determine unique precise rational degrees of belief for many propositions.

of all the coherent probability-functions, the evidential situation rules some out, but leaves more than one not ruled out. Any of the remaining probability-functions would be rational to have in that situation. Rationality, on this view, would not be so restrictive a notion as to make only one epistemic state ideally rational in a given evidential situation.

If one sees the relationship between evidence and rational belief in this way, one might hold that we have no need to model epistemic attitudes other than precise degrees of belief. One might hold that an ideally rational agent is not only permitted, but required, to have one of the probability-functions not ruled out by her evidential situation. On this sort of view, probabilistic coherence would be a straightforward necessary condition on ideally rational beliefs.

I think that there is something attractive about this approach, but it is not obviously correct, for the following reason: Suppose we can make sense of a person's taking a "spread-out" attitude toward a proposition—for example, her confidence in the Republicans winning the next election might best be described as more than 2/5 and less than 3/4, but at no particular place in between those points. Provided that there are such attitudes, one might well think that an ideally rational agent could have them in certain situations. The above spread-out attitude might be thought rational, for instance, if the evidence ruled out only precise degrees of belief outside the 2/5 to 3/4 range.

If having spread-out attitudes can be rational, then we must see the relationship between rational belief and evidence in a more complex way. Spread-out beliefs are naturally represented by *ranges* of precise degrees of belief. Thus, instead of requiring that rational agents adopt particular non-ruled-out precise degrees of belief, we may require that a rational agent's ranges of belief not include ruled-out values. On a liberal version of this view, an agent in the envisaged situation could adopt a precise degree of belief (e.g. 3/5), or a small range (e.g. from 1/2 to 3/5), or a wider range (up to the maximal width of 2/5 to 3/4).

A more restrictive view might hold that agents *may not* adopt precise degrees of belief unless all other degrees of belief are ruled out. This sort of view might require an agent to adopt an attitude corresponding to the widest range of non-ruled-out values. (This view, unlike the others we've looked at, would allow only one ideally rational epistemic response to a given evidential situation.)

I don't want to adjudicate here among the various approaches to this problem. But on either of the last two approaches, a model of ideal rationality should have room for representing spread-out degrees of belief. On these approaches, then, the straightforward use of the probabilistic model will not be sufficient.

Fortunately, it is not difficult to accommodate spread-out beliefs in a way that preserves the intuitive value of the probabilistic model. As various authors have noted, we may represent an agent's belief state not by a single function assigning numbers to propositions, but by a *set* of such functions.[3] The condition of an agent's degree of belief in P being spread out from 0.2 to 0.3 will be represented by her set of belief-functions including members that assign P the numbers from 0.2 to 0.3, but no members assigning P a value outside this range. Instead of requiring that the agent's beliefs be represented by a single probabilistically coherent belief-function, we may require that the agent's beliefs be representable by a set of belief-functions, each of which is probabilistically coherent.[4]

[3] See Kaplan (1996, ch. 1, sect. V) both for an extended argument in support of using sets of probability assignments to represent rational epistemic states, and for references to various implementations of this strategy. Kaplan's version of the RTA is tailored to support his version of this approach.

[4] One might worry that the move to representing agents' attitudes by ranges rather than precise values will not solve the problem. For, insofar as we hold that ideally rational agents are permitted or required to have attitudes toward propositions that are spread out along ranges of degrees of confidence, it might well be thought that those ranges will not typically have precisely determinate boundaries. But the ranges employed by the model do have sharp boundaries. One might even worry that in moving from single values to ranges we've merely traded one instance of misleading precision for two. Now it seems to me fairly plausible that, if rational attitudes toward

This basic idea can be—and indeed has been—filled out in a number of different ways. But each of them preserves the central insight of the simple model we've been examining. On any such view, ideally rational degrees of belief are constrained by the logical structure of the propositions believed, and the constraints are based on the principles of probability. Wherever an agent does have precise degrees of belief, those degrees are constrained by probabilistic coherence in the standard way. Where her credences in certain propositions are spread out, they are still constrained by coherence, albeit in a more subtle way. Thus the normative claim that rationality allows, or even requires, spread-out credences in certain evidential situations does not undermine the basic position that I have been defending: that logic constrains ideal rationality by means of probabilistic conditions on degrees of confidence.

6.2 The Unattainability of Probabilistic Perfection

Although the sort of idealization discussed in the previous section is one source of suspicion about probabilistic models of rationality, another sort of idealization is, I think, significantly more troubling. As many people have pointed out, attaining probabilistic coherence

propositions may be spread out along ranges of degrees of confidence, those ranges themselves will have vague boundaries—there may well be some vagueness in which precise degrees of belief the evidence rules out. But this point is quite compatible with the ranges' providing a vastly improved model of spread-out belief. Consider an analogy: We might represent Lake Champlain as stretching from latitude 43:32:15 N to 45:3:24 N. We would realize, of course, that there really aren't non-vague southernmost and northernmost points to the lake; lakes are objects that lack non-vague boundaries. But representing the lake as ranging between these two latitudes is sufficiently accurate, and vastly better than representing the lake's location by picking some single latitude in between them. Similarly, we might represent an ideally rational agent's attitude toward P in a certain evidential situation as ranging from 0.2 to 0.3. We may well do this while realizing that the lowermost and uppermost bounds on degrees of confidence allowed by the evidential situation are vague. But this representation may yet be very accurate, and a considerable improvement over representing the agent's attitude by a single degree of belief. (Thanks to Mark Kaplan for help on this point.)

is far beyond the capacity of any real human being. Probabilistic coherence, after all, requires having full credence in all logical truths—including complicated theorems that no human has been able to prove. It also places constraints on beliefs about logically contingent matters—constraints that go beyond human capacities to obey. For example, when P entails Q, coherence requires that one's confidence in Q be at least as great as one's confidence in P—even if the entailment is so far from obvious that no human would recognize it. The fact that this sort of "logical omniscience" is built into probabilistic coherence has led many to doubt that coherence can provide any sort of interesting normative constraint on rationality.[5]

Although some have pressed this objection against probabilistic coherence in particular, it is worth noting that it applies equally to the standard idealizations of rational binary belief based on deductive cogency. Deductive closure (and even the very modest closure-under-single-premise-arguments) demands belief in every logical truth; and when P entails Q, closure requires that one believe the latter whenever one believes the former. Consistency forbids belief in contradictory claims, even when—as in the case of Frege's inconsistent foundations for mathematics—the contradiction could elude a brilliant human logician.[6]

In evaluating this general line of objection, it is important to see that one cannot dismiss coherence or cogency as normative ideals merely by pointing out, e.g., that it would seem pretty odd to call Frege "irrational." Acceptance of the ideals in question does not

[5] A related argument is made in Harman (1986, 25–7), which rejects probabilistic methods of belief *revision*, on the grounds that people could not store degree-of-belief information for the number of propositions that would be required for conditionalization. Harman is not mainly interested in synchronic rationality conditions, but the central claim upon which he bases his argument—that we cannot store complete probability distribution information in our heads—would lend itself to an argument parallel to the usual ones based on the unattainability of logical omniscience.

[6] See Cherniak (1986, ch. 4) for a persuasive description of how humans—or any beings remotely like us—are bound to fall far short of deductive logical omniscience.

require this sort of name-calling.[7] When we call someone "irrational," we are saying that he is deficient relative to a contextually appropriate standard, which need not be—and typically is not—the standard of absolute rational perfection. Similarly, when we call someone "immoral," we are (typically) saying something much stronger than that she falls short of absolute moral perfection. And we do not call people "weak" or "stupid" just because they are not as strong or as smart as a being (even a human being) could possibly be.

Accounts of ideal rationality that include logical omniscience are, of course, committed to the claim that Frege's beliefs fell short of perfect rationality. And if one insists on giving an epistemic evaluation of Frege himself (rather than of his beliefs), these accounts of ideally rational belief at least suggest that Frege himself was a less than perfectly rational agent.[8]

But if it seems obviously wrong to call Frege "irrational," it does not seem *obviously* wrong to say that his beliefs (or even Frege himself) fell short of perfect or ideal rationality. It is not an obvious constraint on normative theorizing about rationality that one's account make absolute rational perfection humanly attainable. Thus, the serious worries about the degree of idealization involved

[7] Hawthorne and Bovens (1999, 257) note that those who see failures of logical omniscience as falling short of ideal rationality need not apply the epithet "irrational" to those who fall short of the ideal.

[8] Some defenders of the ideal of probabilistic coherence would reject the implicit connection made above between an agent's *beliefs* meriting rational criticism and the agent *himself* falling short of full rationality. Kaplan writes: "To say that your state of opinion is open to legitimate criticism... is not to say that *you* are open to legitimate criticism... You can hardly be held open to criticism for violations... that are due only to your limited cognitive capacities, limited logical acumen, limited time. Nor can you reasonably be held open to criticism for a violation... that you do not know how to avoid..." (1996, 37–8). In Kaplan's view, an agent's failure to assign a tautology probability 1 may be counted as rational, provided that the failure is excusable owing to the sorts of limitations mentioned above (see Kaplan 2002, 439–40; Armendt 1993, 4 suggests a similar view.) Now it is surely right that there is a sense in which agents cannot rightly be *blamed*, and should not, in Kaplan's phrase, be "called on the carpet," for falling short of standards that it is not within their power to meet. But whether an agent's rationality may be compromised by factors beyond her control is a different question; this will be discussed below.

in cogency- or coherence-based accounts must be formulated more carefully before they can be evaluated. I would like to look at three related but distinct ways in which such worries may be pressed. Although much of what follows would apply to accounts based on deductive cogency as much as it applies to probabilistic-coherence-based accounts, I will confine the discussion to the latter.

6.3 Logical vs Factual Omniscience

L. J. Savage (1967) wondered how his own coherence-based theory of rational decision could be normative, given that it would require one to know—and, on Savage's account, even to risk money on—a proposition about a remote digit of π. Ian Hacking (1967) developed this worry by comparing an ordinary person's lack of such knowledge with an ordinary person's ignorance of various matters of fact. Hacking points out that even mathematical facts are often known by empirical methods, and that these empirical methods may often, for real people, be preferable to strict logical or mathematical proof. Moreover, ordinary rational people may have intermediate degrees of belief in mathematical propositions, just as they have for, say, facts about the locations of subway stops, and may use these degrees of belief in similar ways to make practical decisions. Commenting on Savage's theory of personal probability—which is intended in part to provide a theory of rational belief—Hacking writes: "I do not believe that the theory should acknowledge any distinction between facts found out by *a priori* reasoning and those discovered *a posteriori*" (p. 312).

More recently, Richard Foley has argued that no interesting account of rationality should treat logical omniscience and empirical omniscience differently:

[I]f a logically omniscient perspective ... is an ideal perspective, one to which we aspire and one that we can do a better or worse job of approximating, so too is an empirically omniscient perspective. If this were a reason to regard all departures from logical omniscience as departures from ideal rationality, it would be an equally good reason to regard all departures from empirical omniscience as departures from ideal rationality. But of course, no one wants to assert this. (Foley 1993, 161)

And Philip Kitcher makes a related claim:

Cognitively limited beings, however, can do well or badly in trying to overcome their limitations. We cannot think of them as limited only with respect to "matters of fact"; their perspective on how to proceed in forming their beliefs may also be limited. Thus, just as we excuse ourselves and our predecessors for failure to be omniscient, concepts of rationality and justification *used in assessing the performances of others* should also take into account our methodological foibles.[9] (Kitcher 1992, 67; emphasis in original)

Kitcher cites failures to respect probabilistic rules as an example of the foibles for which limited humans may be excused, in the sense that the foibles would not count against their rationality.

Now it is clearly correct that logical omniscience is no more possible for actual people than is empirical omniscience. But the fact that the two ideals are equally impossible for ordinary humans surely does not by itself suffice to show that failures to attain them should be treated on a par when we are theorizing about *rationality*. The question raised by these arguments is this: from the perspective of theorizing about rationality, should we see failures of logical omniscience as being on a par with failures of empirical omniscience? And it seems to me that a good way to begin thinking about this question is to step back for a moment from thinking directly about omniscience. Let us begin by considering some very ordinary failures of people to believe some very ordinary truths.

Consider the following two cases: Kelly is highly confident that anyone who gets near a grizzly bear cub in the wild is in danger. She is also extremely confident that she is near a grizzly bear cub in the wild. Unfortunately, she doesn't put two and two together, and thus

[9] Goldman (1986, 67–68) argues along these lines as well.

fails to be confident that she's in danger. Nevertheless, she is in danger.

Meanwhile, Cherry is also confident that anyone who gets near a grizzly bear cub in the wild is in danger, and also fails to realize that Kelly is in danger. But the reason that Cherry fails to be confident of this truth is different. Cherry is back at camp, and has no idea that Kelly is near a grizzly cub.

It seems clear here that only Kelly is suffering from a defect in rationality. Kelly's degrees of confidence fail to respect the logical relations among the relevant propositions: Since she's so highly confident of two propositions which together entail a third, she should be very confident of that third proposition.[10] Cherry, on the other hand, is simply missing evidence. This interferes with her ability to know a certain fact. But that by itself has no implications at all for her rationality.

Now it should be noted that there may be cases in which an agent's lack of empirical evidence is itself a manifestation of irrationality—for example, when my fear of hearing bad news prevents me from remembering to check my phone messages. So the point of the example is not that *only* logical lapses count as rational failures. Conversely, the example surely does not in itself show that *all* logical lapses count as rational failures. The point is simply that there is a clear intuitive basis in our ordinary conception of rationality for distinguishing logical lapses from ordinary cases of factual ignorance. And it seems to me that this is just what we should expect. Much of the point, after all, of thinking about rationality is to understand the idea of reasoning well; and reasoning well is not the same thing as being correct. Central to any notion of epistemic rationality is that true beliefs can be held irrationally, and that beliefs held rationally may be false.

[10] Suppose, for example that Kelly has 0.9 credence in the proposition that anyone who gets near a grizzly bear cub in the wild is in danger, and 0.99 credence that she is near a grizzly bear cub in the wild. Her credence in the proposition that she is in danger should be at least 0.88.

Given this motivation for treating ordinary logical lapses and ordinary factual ignorance differently, it is certainly not unnatural to extend the differential treatment to failures of full omniscience. The reason that no one wants to assert that failures of empirical omniscience constitute departures from ideal rationality is simply that, in general, ordinary failures of empirical knowledge do not constitute or even indicate failures of rationality. So, while empirical omniscience and logical omniscience may both be in some sense cognitive or epistemic ideals, they are quite different sorts of ideals. In particular, we have no reason to think that they have equally good (and thus equally bad) claims on being part of ideal rationality.

Now the picture I have been defending is rooted in a distinction—one that underlies our differing common-sense assessments of Kelly and Cherry in the example above—between failures to obtain accurate beliefs because of logical lapses, and failures arising from incomplete evidence. But it is important to see that the importance of the distinction need not be tied to any theoretically rich notion of *a prioricity*. One need not hold, for instance, that real people can achieve infallible or incorrigible beliefs by *a priori* reasoning. One need not hold that there is a clear and sharp distinction between analytic and synthetic sentences. One only needs the basic motivation of characterizing good thinking that doesn't count mere evidential incompleteness as a defect.

Given this motivation, logical omniscience emerges naturally as the limiting case of one of the basic ingredients of good thinking, in a way that empirical omniscience does not. We know that certain structural aspects of the claims we believe have a bearing on their possible truths (e.g. a conjunction is true only if each of its conjuncts is true). Formal logic studies these relationships. It seems clear that many ordinary instances of bad thinking involve failing to respect these relationships (we should not believe a conjunction more strongly than one of its conjuncts). Eliminating this sort of mistake yields, in the limit, logical omniscience. Given that no such result holds for empirical omniscience, it seems to me that we have a clear motivation for treating the two differently when theorizing about ideal rationality.

6.4 Rationality and Deontology

Another way of developing the worry about excess idealization supplements the basic empirical observation—that coherence is humanly unattainable—with a conceptual claim about rationality. Rationality, the thought goes, is a normative notion, and as such must be constrained by the capacities of those to whom it is applied. To say that rationality requires that Rusty give all tautologies maximum credence is to say that (epistemically speaking) Rusty *ought to* give them all maximum credence. But if "ought" implies "can," this last claim will be true only if Rusty has the *capacity* to recognize the tautologies and give them maximum credence—which he clearly does not. Thus, giving maximum credence to all tautologies cannot be a requirement of rationality. We might call the conception of rationality from which this argument springs the deontological concept of rationality.

There has, of course, been much discussion of "ought"-implies-"can" principles in ethics, and in the related literature on free will and determinism. The principle that moral criticism—or at least a central kind of moral criticism—is subject to the condition that the agent "could have done otherwise" is clearly attractive. Of course, this principle has been rejected by many, and even those who maintain it have sometimes found it difficult to understand it in a way that meshes both with our understanding of the way the world works, and with our intuitive moral judgments. But without entering into this complex literature, I think that we can see reason to resist any quick move from a deontological notion of rationality to a rejection of coherence as a rational ideal. To begin with, we should note that it is far from obvious that ideals of moral perfection are subject to the sort of attainability requirement that is being urged for ideals of rational perfection: it does not seem outrageous or silly (or even implausible, I think) to suggest that absolute moral perfection is not psychologically attainable by actual human beings.

Moreover, the unattainability of human moral perfection can even be integrated with an approach to ethics that acknowledges a robust "ought"-implies-"can" principle. The "ought"-implies-"can" principle would of course place limitations on the "oughts" generated by our moral ideal: we might well think that it would be wrong to say that a person "ought" to be morally perfect. But this would not prevent us from taking our moral ideal seriously: we might still assess actions (or agents) morally with the aid of a concept of what moral perfection would be. And these assessments might even be part of a clearly deontological framework. We might, for example, hold that, insofar as it was possible for a given person to come closer to a moral ideal than he has come so far, he ought to do so. Thus, although the moral ideal might be attainable for no one, it might yet play a crucial role in grounding the moral obligations of each agent, obligations that were conditioned by particular facts about what that person could achieve.

Thus, even if we grant the premise that rationality is a deontological notion closely tied to obligation (or duty, deserved praise and blame, or related notions), and even if we grant that the relevant epistemic "oughts" are conditioned by the agent's capacities, it would not undermine taking probabilistic coherence as an aspect of ideal rationality. For as the analogy to morality suggests, we would still have no theoretical reason to insist that an ideal of rational perfection must be attainable by actual people.[11]

[11] As noted above (fn. 8), Kaplan sees any violation of his version of coherence as laying the agent's state of opinion open to rational criticism, but he exempts the agent herself from rational criticism in cases where she could not have been expected to achieve coherence. Kaplan's discussion (2002, 439–40) suggests that the agent herself may be held open to criticism (and called not rational) for violations of coherence that she could easily have avoided (and which are thus not "excused" by factors such as limited logical acumen). On such a treatment of rationality ascriptions *to agents*, a version of coherence plays a role similar to that envisioned for moral ideals in the text. For an example of this sort of model applied to epistemic deontology, see Kornblith (1983).

But should we grant that rationality is a deontological concept which embodies obligations subject to an "ought"-implies-"can" principle? Most discussion of this sort of issue in the literature involves epistemic justification, rather than rationality. It has also been connected less strongly with worries over the unattainability of logical ideals than with worries over our apparent lack of voluntary control over our beliefs. (If one takes epistemic justification to be closely related to some notion of epistemic blamelessness, or to satisfaction of epistemic obligations, one might worry that if our beliefs are not under our voluntary control then we cannot be blamed for believing the way we do, or even have obligations to believe otherwise than we do.)[12] But the discussions provide, I think, additional reasons for doubting that the deontological approach to pressing the ideality worry will succeed.

In addressing the problem with justification, one possibility is to argue that the "oughts" associated with a deontological notion of epistemic justification are not the sort that entail "can." Richard Feldman (2001, 87 ff.) argues that certain "oughts"—he calls them "role oughts"—may be detached from "can." Feldman cites examples such as "Teachers ought to explain things clearly," and points out that this seems true even though some teachers are incapable of explaining things as clearly as they ought to. Feldman writes, "It is our plight to be believers. We ought to do it right. It doesn't matter that in some cases we are unable to do so."[13]

Although I am sympathetic with Feldman's argument that there are senses of "ought" that do not imply "can," and although interpreting epistemic "oughts" this way might assuage the deontological

[12] Plantinga (1993, ch. 1) gives an extensive survey of the deontological thread in epistemology. Alston (1985 and 1988) are classic discussions. And Steup (2001) includes some recent work on the topic.

[13] See Feldman (2001, 88). I should point out that Feldman sees the standards governing at least some roles—such as that of teacher—as being in part constrained by general human capacities. If epistemic oughts were constrained in this way, they presumably could not require probabilistic coherence. I should also note that Feldman rejects a strong epistemic deontology which would *blame* people for violating their epistemic obligations.

version of the excess idealization worry, I do not want to pursue this issue here. For our purposes, the important question is whether epistemic rationality must be seen as deontological in a strong sense that embodies an "ought"-implies-"can" principle. And it seems to me that we have little reason for thinking that it must be thought of in this way.[14]

Let us first turn to epistemic justification. No doubt, there are deontological connotations to "justification," but there are also strong reasons for doubting that the property we are really interested in when we theorize about epistemic justification should be conceived of deontologically (at least in the "ought"-implies-"can" sense). Alston (1985, 1988) divides deontological accounts of justification into two sorts. The first presupposes that belief is under the direct voluntary control of the agent. If this were true, then in many cases the "can" precondition for many epistemic "oughts" would be satisfied. This sort of account fails, however, as a result of the fact that we typically lack direct voluntary control over the vast majority of our beliefs.

The second sort of account Alston considers presupposes only that we have indirect voluntary control over our beliefs, by way of our control over our belief-forming and belief-maintaining activities. Alston uses the analogy of blaming someone for her poor health in a case where her poor health was not directly under her control, but could have been prevented by her doing things (e.g. exercising) that were under her direct voluntary control. (As Alston points out, we could not blame a person for her poor health if it was not preventable by factors within her control.) Similarly, we might

[14] It is worth noting that some of our most central moral concepts may also be non-deontological. Some have worried that "ought"-implies-"can" principles threaten morality, given certain assumptions about the world (e.g. determinism). But a persuasive case can be made for the claim that some of our key moral notions—perhaps good and right—are disentangleable from this sort of deontology. On this conception, we may judge acts (and even agents) morally without supposing that they could have done otherwise. See Pereboom (2001, chs. 5–7) for an extended defense of the claim that a robust morality may be maintained even if one acknowledges that moral agents typically lack the control that would be required for them to do other than they end up doing.

distribute epistemic blame for beliefs that resulted from the agent's failure to meet his epistemic obligations with respect to these indirect belief-controlling factors. But this sort of account fails as well. In many cases, people's beliefs seem blatantly unjustified, even though the people never had control over factors that could have caused them to adopt better beliefs. Alston cites deficiency in cognitive powers and subjection to irresistible (but non-rational) persuasion among the sorts of cases in which, as he puts it, "we could, blamelessly, be believing *p* for outrageously bad reasons" (1985, 96). Insofar as justification is an epistemically valuable state closely related to reasonable belief, then, to say that a person's belief is unjustified, one need not imply that he could have done better.

Alston argues that what we really are interested in when we theorize about epistemic justification is a non-deontological but still clearly *evaluative* notion. (Roughly, for Alston, a belief is justified if it is based on adequate grounds; such believings are good from the epistemic point of view, but no assumption is made that such believings are within the direct or indirect voluntary control of the believer.) And it seems to me that this sort of understanding is even more plausible when applied to epistemic rationality. True, in *some* cases where we call someone's beliefs irrational, we may also think that she could have avoided believing irrationally. For example, we might think that if Kelly had tried harder to keep in mind her rules of wilderness safety, she would have realized that her proximity to that bear cub put her in danger. To the extent that we believe that Kelly had control over these factors of effort and attention, we may even blame her for her irrationally low degree of confidence that she's in danger. But many—perhaps most—cases of irrational belief are not like this at all.

Obvious cases of involuntary irrational beliefs include those that are caused by severe psychological disorders. If I'm quite certain that I must wear an aluminum foil hat to keep the government from reading my thoughts, I have a clearly irrational belief; but this verdict of irrationality does not in any way presuppose that

I could somehow have avoided believing as I do. Psychedelic drugs are also capable of producing irrational beliefs over which the believer does not have control (if one is inclined to mete out epistemic blame for consuming hallucinogens, we may surreptitiously slip the substance to an unwitting gedankenexperimental subject). And we needn't even turn to psychopathology or drugs to find cases of clearly irrational belief that are not obviously subject to the agent's control. Ordinary people often have superstitious beliefs that even they realize are irrational. It is not clear that all such beliefs are outside of the agents' control. But I don't think that we would in the slightest be tempted to withdraw our verdict of epistemic irrationality if we found out that some such agent was psychologically incapable of giving up his superstition.[15]

Thus the kinds of reasons Alston gives for rejecting a deontological account of epistemic justification seem to apply even more clearly to epistemic rationality.[16] Rationality is a good thing, like sanity, or intelligence. It is not the same thing as sanity, or as intelligence; but it is more similar to these notions than it is to notions involving obligation or deserved blame. We may call a person's beliefs irrational without implying that she had the capacity to avoid them. In fact, *pace* Kaplan, we may even call a *person* irrational without implying that she could have done better. In doing so, we are clearly evaluating the person *qua* epistemic agent. We are not holding her open to criticism in any sense that implies that she is to blame for her sorry epistemic state. But not all evaluation need be circumscribed by the abilities of the evaluated. In epistemology, as in various other arenas, we need not grade on effort. And what goes for the harsh-sounding verdict of irrationality

[15] A similar point about unjustified yet involuntary belief is made by Richard Feldman and Earl Conee (1985, 17–19), using the example of a paranoid man. Feldman and Conee argue that their evidentialist account of justified belief does put justification within the reach of normal humans; but they also argue that it would not refute their view if it entailed that normal humans could not avoid unjustified beliefs.

[16] Alston himself (1985, 97–8, fn. 21) indicates that the non-deontological notion that he takes epistemology to be concerned with would better be called by a name other than "justification," with its connotations of obligation and blame.

goes even more clearly for the more moderate verdict of "less than perfectly rational."

There is nothing mysterious about evaluative concepts whose application is not directly constrained by human limitations, even if the evaluations apply to distinctively human activities. To look at just one example, consider goodness of chess play. We can see right away that chess judgments typically abstract from the psychological limitations of individual players. I am a poor chess player, and though I undoubtedly could improve, it's clear that no amount of effort would allow me to achieve chess excellence. If I am unwise enough to play a game of chess and, because of my lack of proficiency, pass up a winning strategy, I am playing less well than someone who, in the same situation, played the winning strategy—despite the fact that I simply could not have come up with that strategy.

Moreover, our fundamental metric of goodness for chess play flows from an ideal that is not even limited by *general* human psychological constraints. True, our ordinary quality judgments about chess players are expressed in terms that are relativized to general human capacities (or, often, even more narrowly relativized to the general capacities of certain sorts of humans, as when we call an eight-year-old child an excellent chess player because she can beat most adults). We would not call Kasparov a "bad chess player" just because he failed to play an available strategy that can be proved—though only through some mathematical analysis far too complex for any human to have performed in the time allotted—to guarantee victory. But underlying all of these relativized judgments is an absolute scale that makes no reference at all to human cognitive limitations. Though we don't blame Kasparov, or call his passing up the winning strategy a "bad play," we will readily acknowledge that playing the victory-guaranteeing strategy would have been better chess. And if a being with superhuman cognitive powers learned to play chess, and came to use such strategies successfully, it would simply be a better chess player than Kasparov, or any human. There are some arenas in which perfection is humanly attainable, and some in which it is not. The considerations above merely suggest that, in this respect, rationality is more like chess than it is like tic-tac-toe.

6.5 Cognitive Advice and the Interest of Epistemic Idealization

A third way of pressing the excess-idealization worry is compatible with seeing a principled difference between logical and factual ignorance, and also does not rely on claiming that rationality has a conceptual connection with any sort of deontological "ought"-implies-"can" principle. Instead, it supplements the initial empirical observation (that probabilistic coherence is humanly unattainable) with a methodological claim about the purpose of theorizing about rationality. The claim is not that epistemology must concern itself with obligation or blame. Rather, it is that the standards or ideals that epistemology invokes must earn their keep by helping us achieve epistemic improvement. Kitcher calls this the "meliorative dimension" in epistemology. He writes:

> if analysis of current concepts of rationality and justification, or delineation of accepted inferential practices, is valuable, it is because a clearer view of what we now accept might enable us to do better. (Kitcher 1992, 64)

A similar sentiment is expressed by Hilary Kornblith; in discussing epistemic ideals, he writes:

> Ideals are meant to play some role in guiding action, and an ideal that took no account of human limitations would thereby lose its capacity to play a constructive action-guiding role.... Epistemic ideals of this sort would fail to make sense of the interest of epistemological theorizing. (Kornblith 2001, 238)

Before evaluating these claims, we should note that there are different ways in which epistemic ideals might be required to yield advice for epistemic improvement. A demanding condition would require that epistemic ideals yield "rules for the direction of the mind" susceptible to self-conscious first-person application. A looser condition might allow for epistemic ideals to be validated by less direct employment in improving our epistemic lot, for example by helping us to come up with advice to educators for improving the cognitive functioning of the young. But for the purposes of the discussion below, I'll focus on the underlying claim that epistemic ideals are of interest only insofar as they can serve the practical end of producing actual epistemic improvement in humans.

As Kitcher (1992, 65) indicates, the demand that epistemology yield usable advice has been emphasized in recent epistemological naturalism. I do not want to take a stand here on whether epistemic ideals such as coherence and deductive cogency compare favorably or unfavorably with naturalistically favored concepts of justification based on, e.g. reliable belief-forming processes, when it comes to playing a helpful advice-giving role in epistemic improvement projects. But I think that Kaplan is correct in holding that the fact that an ideal is not perfectly attainable does not preclude it from playing a regulative role.[17] If this is right, then even if we accepted the claim that melioration is central to epistemology (a claim I want to examine more carefully below), the unattainability of probabilistic perfection would not in itself vitiate coherence as an epistemic ideal.

To see that there is room for unattainable ideals even in the most clearly pragmatic of endeavors, consider the endeavor of designing a car. There are a number of different good-making dimensions along which cars may be evaluated: fuel efficiency, acceleration, handling, safety, etc. Let's concentrate on efficiency. Efficiency seems to enter into the evaluation of car designs in a fairly simple way: the more efficient a car is, the better.[18] Now suppose someone objected to this characterization as follows: "Your evaluative

[17] See Kaplan (1994, 361–362; 2002, 439–40).

[18] Of course, measuring efficiency is itself not completely simple: one car may be efficient at low speeds, another at high speeds, etc. But this point will not affect the argument.

scheme imposes an unrealistic standard. Are you trying to tell me that the Toyota Prius hybrid, at 49 mpg, is an "inefficient" car? On your view, the very best car would use no energy at all! But this is technologically impossible. Indeed, it's even more deeply impossible: the very laws of physics forbid it!" How should we react to such an objection?

To begin with, we needn't accept our objector's invitation to call the Prius "inefficient": instead, we should explain to him that the applicability of such epithets is determined by contextual factors, not directly by comparisons with perfect ideals. But we may cheerfully admit that the Prius isn't perfectly efficient. In fact, we may well grant that achievement of perfect efficiency is deeply impossible. We should also grant that getting anywhere close to this ideal is not possible with anything resembling current technology; indeed, seeking certain high levels of efficiency would be at best a waste of time. Finally, we should grant that, in designing cars, one should not seek to maximize efficiency at all costs: the best car one can design today will probably trade off some efficiency for acceleration.

Now, do any of these concessions vitiate our ideal of efficiency? It seems to me that they do not. Clearly, there's no level of efficiency above which further efficiency would for some reason not be desirable. If the efficiency achievable with our technology maxes out at 517 mpg, we need not despair, or blame ourselves. We might even feel justifiably proud of having designed such an efficient car. But having done that, we should not go on to conclude that the car is ideal. Like so many of our practical improvement projects, new car design depends on evaluations which presuppose values that we know in advance are not maximally realizable. There is nothing paradoxical about this. If the Martians have used special Martian materials to design cars that are as good as ours in other respects, but more efficient, then their cars are better than ours in a perfectly straightforward sense, and we should have no problem acknowledging this.[19]

[19] Of course, even the Martians will be bound by the laws of physics. What if these put some upper limit on fuel efficiency well below 100%—say it worked out to 1517 mpg for cars meeting some standard other specifications? As Kornblith points out

These considerations seem to me to show that even adopting a meliorative conception of epistemology would not preclude our recognizing the normative force of ideals whose perfect—or even nearly perfect—realization was far beyond human capacities. So even if we granted that our theorizing about rationality needed to be tied very closely to our aspirations for the epistemic betterment of ourselves and our fellows, we would not have reason to conclude on that basis that coherence was precluded from being a rational ideal guiding that project.

Moreover—and, I think, more importantly—it is far from clear that epistemology must be tightly tethered to meliorative aspirations. There is, to be sure, historical precedent in epistemology for seeking cognitive improvement; Goldman (1978, 228) cites Descartes and Spinoza, and Kitcher (1992, 64) cites Bacon and Descartes, as antecedents. But the existence of meliorative ambitions in certain great epistemologists does not show these ambitions to be essential ingredients for any interesting epistemology.

At least some reason for doubting the ultimate importance of the meliorative project can be found, I think, in a realistic appraisal of its prospects for success. To put the point bluntly, it is hard to believe that the advice generated by epistemological theorizing is likely to serve as an important (personal or social) force for

(in correspondence), there would be a sense in which a car that achieved that level of efficiency was ideal. I agree that such a car would be ideal in the sense that it would be the most efficient car nomologically possible. But that would not, I think, undermine efficiency's status as an ideal. Let us assume that the laws that impose this limit do not somehow render the notion of greater efficiency incoherent. (If they did, then the 1517 mpg would realize maximum efficiency.) We might now ask the question: what is the basic good-making property for cars? When we evaluate them, are we scoring them on a scale of efficiency, or are we really scoring them on a scale of "efficiency-up-to-1517-mpg"? I can see no reason to insist on the latter interpretation, and can see real reasons for prefering the former. First, the former suggestion has an obvious advantage in simplicity, and connects this dimension of our car evaluations with other efficiency-based evaluations that don't involve any 1517-mpg limit. Also, the former interpretation allows us to make sense of thoughts such as "If the laws of physics were a bit different, we could build cars that were better, because they would be more efficient." Thus, it seems to me that it is efficiency itself that is the good-making feature in car design, and this would remain true even if the limits to our achievements in efficiency were nomological rather than merely technological.

cognitive improvement. The point is not that cognitive improvement is impossible; in fact, I suspect that quite the opposite is true. Most obviously, psychology can help us improve our cognitive skills, either directly (as when I read research on memory improvement, and use it to improve my own memory) or indirectly (as when psychological research on learning informs pedagogy). Courses in statistics, logic, or experimental design can improve students' thinking in many important contexts. And studying history, microbiology, French, music, anthropology, number theory, philosophy, etc., all contribute not just to knowledge of the subject matter, but to general mental improvement. Less direct approaches to one's own cognitive improvement may include exercising, playing chess, sleeping enough at night, and drinking enough coffee during the day. One may promote cognitive improvement in one's children by talking and reading to them, and a society may promote cognitive improvement more widely by establishing a free breakfast program in its public schools. The vast variety of meliorative methods is only hinted at in the above list. But it seems to me that the list suffices to make a point: a frank comparative assessment of the potential contribution that epistemological theorizing has to make among these strategies would reveal it to play, at best, a relatively minor role.

One might object that epistemologists still play a crucial role of choosing our epistemic objectives, while other researchers merely devise means to those objectives. Of course, when there is controversy about the efficacy of some particular strategy for cognitive improvement ("Will playing Mozart for my fetus increase her IQ?") it is primarily the psychologists to whom we turn. But mustn't the psychologists depend on some specification of what cognitive improvement consists in?

It is true that any practical work on cognitive improvement depends on some notion of what such improvement would consist in. But it seems to me that the goals that we care about achieving are generally quite obvious and commonly accepted. Kitcher writes:

Say that an agent's formation of a belief is *externally ideal* just in case that belief was generated by a process that, among all the processes available to the agent in his context, was of a type whose expected epistemic utility was highest. Here the notion of expected epistemic utility is parasitic on an account of cognitive goals and on an assignment of frequencies of success within a contextually determined class of situations. The meliorative project is to identify processes that are externally ideal. (1992, 66)

But if one thinks about goal-setting aspects of this project that go beyond platitudes such as "believing truths is good" and "believing falsehoods is bad," one is left with questions such as "which sorts of truths are most important to believe?" or "how do we individuate the processes (or define relevant contexts) for the purposes of epistemic utility calculations?" These questions may be interesting to the philosopher concerned with epistemic justification. But it is hard to see general philosophical answers to such questions as playing a significant practical role in guiding our individual or societal attempts at cognitive improvement. And when one thinks about other questions that have occupied epistemologists—even naturalistic ones—recently (whether animals have knowledge; or whether clairvoyants, brains in vats, victims of brain tumors, or tourists in fake-barn zones have rational or justified beliefs), the potential payoffs in practical advice are even harder to discern.[20]

Nevertheless, I do not want to rest too much importance on denying epistemology's potential for practical cognitive payoff. For even if my doubts are misplaced, the argument against the interest of unattainable ideals depends on a claim much stronger than the claim that epistemologists' advice has an important role to play in furthering our cognitive improvement. The argument depends on the claim that there is *no* interesting project in epistemology whose interest is independent of its potential to generate cognitive advice. And this sort of claim seems far more doubtful. If philosophy legitimately studies the nature of truth, the question of scientific realism, the distinction between primary and secondary qualities,

[20] For an extended argument against the importance of cognitive advice-giving by epistemologists, see Foley (1992 or 1993, sect. 3.3).

the relation between mind and body, the semantics for propositional attitude ascriptions, and the controversy between cognitivist and non-cognitivist analyses of moral utterances, then it cannot be any kind of general requirement on interesting philosophy that it yield usable practical advice.

Can we not simply be interested in the nature of rationality for its own sake, whether or not our learning about this nature is likely to help us become better thinkers? A possible reason for denying this is suggested by Kitcher. Suppose that, in reply to the complaint that probabilistic coherence sets an unachievable standard, one claims that it nevertheless is constitutive of ideal rationality. In the version of this reply that Kitcher considers, this claim is made on the basis of conceptual truth or analyticity. Kitcher writes of such replies:

> an appropriate challenge is always, "But why should we care about these concepts of justification and rationality?" The root issue will always be whether the methods recommended by the theory are well adapted for the attainment of our epistemic ends, and that cannot be settled by appealing to our current concepts. (Kitcher 1992, 63–4)

And Stephen Stich (1990, ch. 4) makes a more radical argument along similar lines against the idea that one might value rationality for its own sake. He points out that other cultures may have somewhat similar yet distinct systems of cognitive evaluation, and asks "why one would much care that a cognitive process one was thinking of invoking (or renouncing) accords with the set of evaluative notions that prevail in the society into which one happened to be born" (p. 94). Stich even rejects (on similar grounds) the goal of believing truths. "[I]t's hard to see," he writes, "why anyone but an epistemic chauvinist would be much concerned about rationality or truth" (pp. 134–5). He urges instead that cognitive systems be assessed purely pragmatically, by their likelihood of advancing whatever aims their possessors may have (Stich offers health, happiness, and the well-being of the agent's children as typical examples).

It seems to me that several points need to be made about this sort of argument. First, the claim that, e.g., probabilistic coherence is constitutive of rationality need not depend on any notion of analyticity. Of course, in the philosophical investigation of rationality, we will have to utilize our concept of rationality—for instance, in classifying examples. But in this respect, our investigation of rationality is no different from anyone's investigation of anything. Even in cases of straightforward scientific investigations of the natural world, we rely on our concepts in this way: one cannot pursue ichthyology without having any idea of which organisms are fish. This does not render our scientific investigations mere conceptual analysis; after all, we may even discover that some intuitively correct applications of a concept are mistaken—as in the case of our having taken whales to be fish. Similarly, it seems to me that any epistemologist will rely on our concept of rationality to some extent, e.g. in rejecting the proposal that rational beliefs include all those that make one feel sad, or only those that aren't about turtles. So while Kitcher is certainly right in saying that appeal to concepts will not settle certain means–ends questions, this does not preclude appeals to the concept of rationality from playing an important role in our investigation.

It is also worth pointing out that there is a difference between being interested in finding out which cognitive processes one's society approves of (or refers to by its word 'rational'), and being interested in rationality for its own sake. One may begin with an inchoate understanding of rationality, and want to understand it better. One will certainly make use of one's concept of rationality to help distinguish what one is interested in from other dimensions along which cognitive states and processes may be evaluated. But what *makes* rationality interesting need not be that one's society approves of it (or, certainly, that it is the referent of a particular word).

The legitimacy of using our concept of rationality in epistemology may be underlined, I think, by thinking about Stich's and Kitcher's positive suggestions for how epistemology should proceed. Let us first consider Stich's more radical proposal that we "assess cognitive systems by their likelihood of leading to what their users value" (136). As Kornblith points out, there are many sorts of evaluations, and although the sort that Stich proposes is directed at cognitive states and systems, there does not seem to be anything distinctively *epistemic* about it. Kornblith writes:

> If I could assure world peace by committing some epistemic impropriety, surely it would be worth the price. By identifying epistemic propriety with all-things-considered judgments, Stich makes this thought self-contradictory.... It is hard to see how evaluation relative to [pragmatic] concerns is rightly termed epistemic."[21] (Kornblith 1993, 368–9)

Thus Kornblith, to my mind correctly, rejects Stich's proposal as amounting to eliminativism about epistemic evaluation.

Now it seems to me that there is a lesson implicit in Kornblith's argument that goes beyond the rejection of Stich's proposal. To see this, consider a response that would be natural for Stich to make: "Why should we care whether a proposed system of evaluation fits our culture's definition of 'epistemic'? Only chauvinism could justify such an arbitrary restriction!" This response would, I think, derive from a correct perception about Kornblith's argument: that it relies, at bottom, on an appeal to our concept of the epistemic—a concept on which it is not self-contradictory for a cognition to be in an agent's best interest overall while being epistemically sub-par. The argument criticizes Stich's proposal not as being pernicious, but as failing to address our interest in *epistemic* evaluation. This is not, I think, a defect in Kornblith's argument. It simply highlights the legitimate role of conceptual considerations in doing philosophy.[22]

[21] Goldman (1991, 192–3) makes a parallel point. Although I will discuss only this line of criticism of Stich, there are others. Stephen Jacobson (1992) develops several, including the point that Stich's argument against valuing truth and rationality intrinsically would seem to apply equally to, e.g., health.

[22] I do not mean to imply here that Kornblith would endorse my interpretation of his argument, or the lesson I draw from it. In fact, the paper from which this argument is drawn expresses sympathy for Stich's rejection of conceptually based approaches to understanding epistemic justification.

With this thought in mind, let's turn to Kitcher's more moderate proposal that conceptual concerns be displaced by questions about "whether the methods recommended by the theory are well adapted for the attainment of our epistemic ends." Although Kitcher's proposal avoids Kornblith's charge of changing the subject completely away from epistemology, it seems to me that a closely related objection may be raised. For it seems clear that the fact that a certain method of belief formation is conducive to our epistemic ends—in the long run, given general facts about human psychology and the conditions in which we typically find ourselves—does not settle the question of whether beliefs formed by that method are epistemically rational. To see this, suppose it turns out that those who systematically overestimate their own intelligence, attractiveness, and future prospects are actually more successful in life than those whose self-assessments are more accurate. And suppose that this increased level of success is actually caused by the overblown self-assessments, and that the success includes general success in epistemic endeavors such as science, or history, or just attaining ordinary knowledge of the world.[23] If that is the case, the psychological mechanism responsible for the overblown self-assessments would certainly seem to be well adapted for the attainment of our epistemic ends. But it seems to me that this would hardly show that the distorted beliefs about one's self produced by the mechanism were epistemically rational.[24]

[23] The envisioned possibility is not even far-fetched. In *Positive Illusions: Creative Self-Deception and the Healthy Mind*, S. E. Taylor (1989) presents strong evidence correlating unrealistically high self-assessments with pragmatic success. If bloated self-images could increase energy and positive attitude, it is easy to see how they could promote overall epistemic success.

[24] See Firth (1981) for a parallel point. I should note that nothing above is meant to deny that it might be rational in the pragmatic sense to cultivate self-aggrandizing beliefs. It might even be pragmatically rational to do this if one's practical goals were restricted to, e.g., maximizing one's confidence in true claims and minimizing one's confidence in false ones. Thus, if we discovered effective techniques for promoting overblown self-assessments (say, smiling into the mirror while repeating "Damn, I'm good!!"), these techniques might rightly be recommended by someone whose main concern was with epistemic melioration. But this point only highlights the distinction between the philosophical study of epistemic rationality and the project of general cognitive improvement.

Now one could object that I've given Kitcher's words too simple an interpretation. What's needed, one might claim, is a more carefully nuanced notion of the relation between the belief-regulating process and the beliefs it produces. We could, for example, require that the favored processes reliably produce true beliefs *directly*. Since the attainment of our epistemic goals achieved by inducing bloated self-assessments would be achieved only indirectly, we could on this basis withhold from this process (and the beliefs directly produced thereby) our seal of epistemic approval.

But however reasonable such an objection would be, it raises a crucial methodological question: what is the motivation for insisting on such a refinement? It certainly would not derive from our generalized desire for cognitive improvement—that may be served equally well by direct or indirect methods. It seems clear that the only motivation for such a move would derive from the fact that mechanisms of the sort described above blatantly fail to answer to our concept of epistemic rationality. So, while I would certainly be sympathetic to molding our account of rationality to accommodate counterexamples of the envisioned sort, I would argue that our motivation for doing so reflects our interest in understanding the nature of *rationality*, an interest that is distinct from our general interest in improving the lots—even the epistemic lots—of our fellows or ourselves.

What, then, of the challenge to explain why we should care about our concept of rationality? Part of the answer, it seems to me, is that what we care about is rationality itself, not our concept (or word) per se. But we might still ask ourselves: why should we care about rationality? Shouldn't our account of epistemic rationality provide us with an answer to this question?

In one sense of the question, I think that the answer is "yes." It might have turned out, after all, that our investigation of epistemic rationality did not reveal any interesting principles or ideals. In that case, the correct conclusion to reach might have been that our concept of rationality was just a cultural artifact, and that rationality itself was an arbitrary property, its contours of purely parochial interest. So we have, I think, no guarantee before beginning the investigation that we should be seriously interested in rationality, whatever it turns out to be; the interest of the project of understanding rationality is in part contingent on its fruits.

Fortunately, I would argue, the project is bearing fruit. Although deductive cogency fails, in the end, to provide constraints on an epistemically important kind of belief, probabilistic coherence fares better. There seems to be little that is provincial or chauvinistic about our interest in the ideal of coherence. It provides a powerful, simple, and intrinsically appealing condition on graded belief. And as the DBA and RTA reveal, it is tied in interesting ways to practical rationality. So in one sense the challenge to explain why we should care about rationality is legitimate, but it is also answerable.

Still, this sort of answer fails to address another version of the "why should we care" question. The question might be interpreted as a demand that our account provide some sort of reason for every agent, or almost every agent, to care about epistemic rationality (presumably, to care about being rational, and thereby to care about understanding the nature of rationality). To put it another way, the question might be formulated as a demand for grounding the norm of rationality in something external to it, in a way that would appeal (or should appeal) to (almost) anyone.[25]

On this interpretation of the challenge, I see less reason to think that it can be given a satisfactory answer of the sort that those raising the question typically seem to want. As we have seen, we have no reason to suppose that epistemic rationality should always turn out to be in one's general practical interest, all things considered. Many types of situation may favor agents with irrational beliefs. A classic example is suggested by Pascal's wager: an agent could be rewarded immensely by a god for adopting a belief that was not supported by the agent's evidence. Kornblith's world peace example makes the same point without restricting the agent's interests to selfish ones. And more realistically, it seems clear that

[25] Kornblith (1993) poses a question along these lines, and offers a grounding in hypothetical imperatives.

it would in no way be paradoxical if psychologists were to discover that certain sorts of irrationality were conducive to general success in life. So the norm of epistemic rationality should not be expected to drop out as a special case of the norm of pragmatic rationality.

Does this concession—that epistemic rationality need not always serve our practical ends, all things considered—somehow suffice to deprive epistemic rationality of all interest or normative force? I see little reason to think so. After all, aesthetic and moral norms seem, on most accounts anyway, to be in the same boat. There is no obvious reason to deny the existence of multiple norms or values, none of which reduces to any of the others. But once we countenance a multiplicity of values or norms that are independent in this sense, it is hard to see any reason for thinking that epistemic norms must flow from non-epistemic ones. Thus, I see no reason for thinking that our interest in epistemic rationality needs to be grounded in seeing it instrumentally, as a mere means to some other, intrinsically valuable, end.

Of course, insofar as there are purely *epistemic* reasons, we may all automatically have epistemic reasons to be epistemically rational. But this seems tautologous. It is like saying we all have moral reasons to be morally good—it will be unsatisfying to someone who seeks to ground the norm in something external to it. Nevertheless, just as one may reject the demand that morality be grounded in self-interest or other non-moral values, one may reject the demand that the interest or value of epistemic rationality be externally grounded. I think we should reject this demand. One may, after all, be interested in epistemic rationality, and one may value epistemic rationality, for its own sake.

6.6 Epistemic Ideals and Human Imperfection

We have seen, then, no reason to presuppose that ideals for epistemic rationality need to be constrained by human cognitive capacities. And, in closing, it seems to me that there is also good reason not to make any such presupposition: it would tend to foreclose certain interesting questions one might have about us, and our relation to the world. In thinking about our beliefs—our chief means of representing the world to ourselves—one might well want to ask questions about how well, in general, these representations operate. There seems to be nothing wrong with this sort of question. We see pretty clearly that we're better at representing the world than chimps, and we can see that some people are better at it than others. If some of the differences we see turn out to lie on a scale the top of which is beyond our reach, this seems like an interesting result, not a defect in the scale.

Consider one more time Kelly, who has high degrees of belief both that anyone near a bear cub in the wild is in danger, and that she is near a bear cub in the wild, but fails to have a high degree of belief that she's in danger. We may compare her with Mark, who, being highly confident that Kelly is near a bear cub in the wild (and that anyone near a bear cub in the wild is in danger), believes strongly that Kelly is in danger. Mark is (all else being equal) more rational than Kelly: his degrees of belief fit together in a way that respects the logical interconnections among the claims believed. And this is so even if, owing to her psychological makeup, Kelly is incapable of doing better cognitively.

But this is not to say that Mark is perfectly rational. There may be some more subtle logical connections among his beliefs that he is not respecting. In fact, given that Mark is human, this is surely the case, even if Mark is as logical-law-abiding as a human could possibly be. But now consider a slightly superhuman being, one with cognitive capacities just a bit greater than Mark's, who respects some of the logical connections that Mark does not. The difference between this creature and Mark would seem to be of exactly the same sort as the difference between Mark and Kelly; and if that is so, there is no reason to deny that the creature is a bit more rational than Mark. It seems, then, that we need not take ourselves, in any simple way, as the measure of all things. We should accept with good grace that the limits of good human thinking need not be the limits of goodness for all thinking.

Furthermore, philosophy in general, and epistemology in particular, need not be directed toward external practical ends. We surely may philosophize because we hope (perhaps optimistically) to help people improve themselves cognitively. But just as surely, epistemologists need not restrict their efforts to improving our educational system, or to producing popular manuals for cognitive self-help. We may philosophize because we want a better understanding of ourselves—of our cognitive natures and our situation in the world. We may philosophize because we want a better understanding of rationality itself. There is plenty of room for questions on these topics to be asked, and plenty of room for a theory of ideal rationality designed to help answer them. If the arguments considered above are correct, then logic, once ensconced in its rightful place as a constraint on ideally rational degrees of belief, provides an important ingredient for these answers.

REFERENCES

Adler, J. (2002), *Belief's own Ethics* (Cambridge, Mass.: MIT).
Alston, W. P. (1985), "Concepts of Epistemic Justification," in Alston (1989).
—— (1988), "The Deontological Conception of Epistemic Justification," in Alston (1989).
—— (1989), *Epistemic Justification: Essays in the Theory of Knowledge* (Ithaca, NY: Cornell University Press).
Armendt, B. (1993), "Dutch Books, Additivity, and Utility Theory," *Philosophical Topics* 21 (1): 1–20.
Bender, J. W. (ed.) (1989), *The Current State of the Coherence Theory: Critical Essays on the Epistemic Theories of Keith Lehrer and Laurence BonJour* (Dordrecht: Kluwer).
BonJour, L. (1985), *The Structure of Empirical Knowledge* (Cambridge, Mass.: Harvard University Press).
Chan, S. (1999), "Standing Emotions," *Southern Journal of Philosophy* 37: 495–513.
Cherniak, C. (1986), *Minimal Rationality* (Cambridge, Mass.: MIT).
Christensen, D. (1996), "Dutch-Book Arguments Depragmatized: Epistemic Consistency for Partial Believers," *Journal of Philosophy* 93: 450–79.
—— (2001), "Preference-Based Arguments for Probabilism," *Philosophy of Science* 68: 356–76.
Cohen, S. (1988), "How to Be a Fallibilist," *Philosophical Perspectives* 2: 91–123.
Cottingham, J. *et al.* (1984), *The Philosophical Writings of Descartes*, vol. 2 (New York: Cambridge University Press).
de Finetti, B. (1937), "Foresight: its Logical Laws, its Subjective Sources," in H. E. Kyburg and H. E. Smokler (eds.), *Studies in Subjective Probability* (Huntington, NY: Robert E. Krieger, 1980).
—— (1977), "Probability: Beware of Falsifications!" in H. E. Kyburg, and H. E. Smokler (eds.), *Studies in Subjective Probability* (Huntington, NY: Robert E. Krieger, 1980).

DeRose, K. (1996), "Knowledge, Assertion, and Lotteries," *Australasian Journal of Philosophy* 74: 568–80.

Eells, E. (1982), *Rational Decision and Causality* (New York: Cambridge University Press).

Evnine, S. J. (1999), "Believing Conjunctions," *Synthèse* 118: 201–27.

——(2001), "Learning from One's Mistakes: Epistemic Modesty and the Nature of Belief," *Pacific Philosophical Quarterly* 82: 157–177.

Fallis, D. (2003), "Attitudes toward Epistemic Risk and the Value of Experiments," to be published.

Field, H. (1977), "Logic, Meaning, and Conceptual Role," *Journal of Philosophy* 74: 379–409.

Feldman, R. (2001), "Voluntary Belief and Epistemic Evaluation," in Steup (2001).

——and Conee, E. (1985), "Evidentialism," *Philosophical Studies* 48: 15–34.

Firth, R. (1981), "Epistemic Merit, Intrinsic and Instrumental," *Proceedings and Addresses of the American Philosophical Association* 55: 5–23.

Foley, R. (1987), *The Theory of Epistemic Rationality* (Cambridge, Mass.: Harvard University Press).

——(1992), "What Am I to Believe?" in S. Wagner and R. Warner (eds.), *Beyond Physicalism and Naturalism* (South Bend, Ind.: University of Notre Dame Press).

——(1993), *Working without a Net* (New York: Oxford University Press).

Goldman, A. (1978), "Epistemics: the Regulative Theory of Cognition," in Kornblith (1985).

——(1986), *Epistemology and Cognition* (Cambridge, Mass.: Harvard University Press).

——(1991), "Stephen P. Stich: *The Fragmentation of Reason*," *Philosophy and Phenomenological Research* 51: 189–93.

Good, I. J. (1962), "Subjective Probability as the Measure of a Non-measurable Set," in his *Good Thinking: The Foundations of Probability and its Applications* (Minneapolis: University of Minnesota Press, 1983).

Hacking, I. (1967), "Slightly More Realistic Personal Probability," *Philosophy of Science* 34: 311–25.

Harman, G. (1970), "Induction: A Discussion of the Relevance of the Theory of Knowledge to the Theory of Induction (with a Digression to the Effect that Neither Deductive Logic nor the Probability Calculus has Anything To Do with Inference)," in M. Swain (ed.), *Induction, Acceptance, and Rational Belief* (Dordrecht: D. Reidel).

—— (1986), *Change in View* (Cambridge, Mass.: MIT).

Hawthorne, J. (1998), "On the Logic of Nonmonotonic Conditionals and Conditional Probabilities: Predicate Logic," *Journal of Philosophical Logic* 27: 1–34.

—— and Bovens, L. (1999), "The Preface, the Lottery, and the Logic of Belief," *Mind* 108: 241–65.

Hempel, C. G. (1960), "Inductive Inconsistencies," reprinted in his *Aspects of Scientific Explanation* (New York: Free Press, 1965).

Horwich, P. (1982), *Probability and Evidence* (New York: Cambridge University Press).

Howson, C. and Franklin, A. (1994), "Bayesian Conditionalization and Probability Kinematics," *British Journal for the Philosophy of Science* 45: 451–66.

Howson, C. and Urbach, P. (1989), *Scientific Reasoning: the Bayesian Approach* (La Salle, Ill.: Open Court).

Hrobjartsson, A. and Gotzsche, P. C. (2001), "Is the Placebo Powerless? An Analysis of Clinical Trials Comparing Placebo Treatment with No Treatment," *New England Journal of Medicine* 344: 1594–1602.

Jacobson, S. (1992), "In Defense of Truth and Rationality," *Pacific Philosophical Quarterly* 73: 335–46.

Jeffrey, R. (1965*a*), *The Logic of Decision* (Chicago: University of Chicago Press, 2nd edn. 1983).

—— (1965*b*), "New Foundations for Bayesian Decision Theory," in his *Probability and the Art of Judgement* (New York: Cambridge University Press, 1992).

—— (1970), "Dracula meets Wolfman: Acceptance vs Partial Belief," in Swain (1970).

—— (1991), "Introduction: Radical Probabilism," in his *Probability and the Art of Judgement* (New York: Cambridge University Press, 1992).

Joyce, J. M. (1998), "A Nonpragmatic Vindication of Probabilism," *Philosophy of Science* 65: 575–603.

Kaplan, M. (1994), "Epistemology Denatured," in P. A. French, T. E. Uehling Jr, and H. K. Wettstein (eds.), *Midwest Studies in Philosophy*, vol. XIX: *Philosophical Naturalism* (South Bend, Ind.: University of Notre Dame Press).

—— (1996), *Decision Theory as Philosophy* (New York: Cambridge University Press).

—— (2002), "Decision Theory and Epistemology," in *The Oxford Handbook of Epistemology* (New York: Oxford University Press).

KITCHER, P. (1992), "The Naturalists Return," *Philosophical Review* 101: 53–114.

KLEIN, P. (1985), "The Virtues of Inconsistency," *The Monist* 68: 105–135.

KORNBLITH, H. (1983), "Justified Belief and Epistemically Responsible Action," *Philosophical Review* 92: 33–48.

——(ed.) (1985), *Naturalizing Epistemology*, 1st. edn., (Cambridge, Mass.: MIT).

——(1993), "Epistemic Normativity," *Synthèse* 94: 357–376.

——(2001), "Epistemic Obligations and the Possibility of Internalism," in A. Fairweather and L. Zagzebski (eds.), *Virtue Epistemology: Essays on Epistemic Virtue and Responsibility* (New York: Oxford University Press).

KYBURG, H. (1970), "Conjunctivitis," in Swain (1970).

LEHRER, K. (1974), *Knowledge* (New York: Oxford University Press).

——(1975), "Reason and Consistency," reprinted in his *Metamind*, (New York: Oxford University Press, 1990).

LEVI, I. (1967), *Gambling with Truth* (New York: Alfred A. Knopf).

——(1991), *The Fixation of Belief and its Undoing: Changing Beliefs through Inquiry* (New York: Cambridge University Press).

MAHER, P. (1993), *Betting on Theories* (New York: Cambridge University Press).

——(1997), "Depragmatized Dutch Book Arguments," *Philosophy of Science* 64: 291–305.

——(2002), "Joyce's Argument for Probabilism," *Philosophy of Science* 69; 73–81.

MAKINSON, D. C. (1965), "The Paradox of the Preface," *Analysis* 25: 205–7.

MELLOR, D. H. (1971), *The Matter of Chance* (Cambridge: Cambridge University Press).

NELKIN, D. (2000), "The Lottery Paradox, Knowledge, and Rationality," *Philosophical Review* 109: 373–409.

NOZICK, R. (1993) *The Nature of Rationality* (Princeton: Princeton University Press).

PEREBOOM, D. (1991), "Why a Scientific Realist Cannot Be a Functionalist", *Synthèse* 88: 341–58.

——(1994), "Stoic Psychotherapy in Descartes and Spinoza," *Faith and Philosophy* 11: 592–625.

——(2001), *Living without Free Will* (New York: Cambridge University Press).

PLANTINGA, A. (1993), *Warrant: the Current Debate* (New York: Oxford University Press).
POLLOCK, J, (1983), "Epistemology and Probability," *Philosophy of Science* 55: 231–52.
—— (1986), "The Paradox of the Preface," *Philosophy of Science* 53: 246–58.
—— (1990), *Nomic Probability and the Foundations of Induction* (New York: Oxford University Press).
—— and CRUZ, J. (1999), *Contemporary Theories of Knowledge* (New York: Rowman & Littlefield).
POPPER, K. (1959) *The Logic of Scientific Discovery* (New York: Hutchinson).
RAMSEY, F. P. (1926), "Truth and Probability," in H. E. Kyburg, and H. E. Smokler (eds.), *Studies in Subjective Probability* (Huntington, NY: Robert E. Krieger, 1980).
ROORDA, J. (1997), "Fallibilism, Ambivalence, and Belief," *Journal of Philosophy* 94: 126–55.
RYAN, S. (1991), "The Preface Paradox," *Philosophical Studies* 64: 293–307.
—— (1996), "The Epistemic Virtues of Consistency," *Synthèse* 109: 121–41.
SAVAGE, L. J. (1954), *The Foundations of Statistics* (New York: John Wiley).
—— (1967), "Difficulties in the Theory of Personal Probability," *Philosophy of Science* 34: 305–10.
SKYRMS, B. (1975), *Choice and Chance*, 2nd edn., (Encino, Cal.: Dickenson).
—— (1980), "Higher Order Degrees of Belief," in D.H. Mellor (ed.), *Prospects for Pragmatism* (New York: Cambridge University Press).
—— (1990), *The Dynamics of Rational Deliberation.* (Cambridge, Mass.: Harvard University Press.
STEUP, M. (ed.), (2001), *Knowledge, Truth, and Duty* (New York: Oxford University Press).
STICH, S. (1990), *The Fragmentation of Reason* (Cambridge, Mass.: MIT).
SWAIN, M. (ed.), (1970), *Induction, Acceptance, and Rational Belief* (New York: Humanities Press).
TAYLOR, S. E. (1989), *Positive Illusions: Creative Self-Deception and the Healthy Mind* (New York: Basic Books).
UNGER, P. (1975), *Ignorance: A Case for Scepticism* (New York: Oxford University Press; reissued in 2002).
VAN FRAASSEN, B. (1995), "Fine-grained Opinion, Probability, and the Logic of Full Belief," *Journal of Philosophical Logic* 24: 349–77.
VOGEL, J. (1990), "Are There Counterexamples to the Closure Principle?" in M. D. Roth and G. Ross (eds.), *Doubting: Contemporary Approaches to Skepticism* (Dordrecht: Kluwer).

VOGEL, J. (1999), "The New Relevant Alternatives Theory," *Philosophical Perspectives* 13, 155–180.

WEINTRAUB, R. (2001), "The Lottery: A Paradox Regained and Resolved," *Synthèse* 129: 439–49.

WILLIAMSON, T. (1996), "Knowing and Asserting," *Philosophical Review* 105: 489–523.

—— (2000), *Knowledge and its Limits* (New York: Oxford University Press).

ZYNDA, L. (2000), "Representation Theorems and Realism about Degrees of Belief," *Philosophy of Science* 67: 45–69.

INDEX

Printed in the USA/Agawam, MA
March 22, 2011
557273.055